live green, Calgary!

LOCAL PROGRAMS,
PRODUCTS & SERVICES
TO GREEN YOUR LIFE
AND SAVE YOU MONEY

LAUREN MARIS

RMB
Victoria Vancouver Calgary

Copyright © 2009 Lauren Maris

All rights reserved. No part of this publication may be reproduced, stored in a retrieval system, or transmitted in any form or by any means—electronic, mechanical, audio recording, or otherwise—without the written permission of the publisher or a photocopying licence from Access Copyright, Toronto, Canada.

Rocky Mountain Books
#108 – 17665 66A Avenue
Surrey, BC V3S 2A7
www.rmbooks.com

Rocky Mountain Books
PO Box 468
Custer, WA
98240-0468

Library and Archives Canada Cataloguing in Publication

Maris, Lauren

  Live green, Calgary! : local programs, products and services to green your life and save you money / Lauren Maris. — 1st Rocky Mountain Books ed.

ISBN 978-1-897522-57-8

  1. Sustainable living—Alberta—Calgary—Handbooks, manuals, etc. 2. Sustainable living—Handbooks, manuals, etc. 3. Green products—Alberta—Calgary—Directories. 4. Environmental protection—Alberta—Calgary Citizen participation. I. Title.

GE199.C3M37 2009   640   C2009-903728-9

Printed in Canada

Rocky Mountain Books acknowledges the financial support for its publishing program from the Government of Canada through the Book Publishing Industry Development Program (BPIDP), Canada Council for the Arts, and the province of British Columbia through the British Columbia Arts Council and the Book Publishing Tax Credit.

The interior pages of this book have been produced on 100% post-consumer recycled paper, processed chlorine free and printed with vegetable-based dyes.

This book is intended to provide options for more environmentally responsible living in Calgary. However, available programs, products and services change rapidly. The author cannot guarantee that all links or contact information will remain correct for the life of this edition of *Live Green, Calgary!* As well, not every service that identifies itself as green is included in this guide. If you know of a local green program, product or service that might qualify for future versions of *Live Green, Calgary!* or if you would like to let me know about current information regarding the information contained in this version of the guide, please contact me at info@livegreencalgary.com.
No individual or organization has paid to appear in this book.
403-896-5096
www.livegreencalgary.com

There are no passengers on spaceship earth. We are all crew.

— Marshall McLuhan,
Canadian communications theorist,
educator, philosopher (1911–1980)

CONTENTS

	Introduction	7
1	Home Heating	11
2	Indoor Water Use	17
3	Electricity	22
4	Household Cleaning	28
5	Yard and Garden	31
6	Home Building and Renovation	38
7	Home Décor	47
8	Transportation	52
9	Food and Drink	58
10	Waste and Recycling	67
11	Children	77
12	Clothing and Personal Care	83
13	Entertainment and Recreation	89
14	Get Greener in the Workplace	95
15	General Resources for Living Green	104
	Appendix	111

INTRODUCTION

Since the release of the first edition of *Live Green, Calgary!* in 2007, I have spoken with hundreds of Calgarians about living more environmentally responsible lifestyles. After making a presentation at a major oil company I told the company representative that I had worked in oil and gas for nearly a decade. She looked at me with great relief and said, "Oh, why didn't you say that you're one of us?"

I have been interested to discover in the past several years that "eco-shame" is everywhere in Calgary. When I ask someone if they perform a particular environmentally responsible act like composting, they often sheepishly say no, that they know they should but they don't. They look ashamed as they make excuses for their behaviour. I am an eco-sinner myself. My car is not the most fuel-efficient vehicle on the road, and I find myself making excuses for it before anyone has a chance to say anything about it. When I go shopping, I sometimes wonder if someone is going to jump out from behind a car in the parking lot and reprimand me for my 10-litre-per-100-kilometre gas-guzzler.

We all have eco-sinned. The trick to moving forward and "green"ing our lives is not to be paralyzed by eco-shame. No, I don't currently drive one of Transport Canada's most fuel-efficient vehicles, but when I do buy another vehicle you can be darned sure that its fuel efficiency will be top priority. In the meantime, I have made other significant changes in my life. I got my vermicomposting set-up going (thanks for sharing your worms, Lisa!) and have reduced my waste to landfill by a third. I have installed a low-flow showerhead (which has exceeded expectations for performance, by the way) and have been working on converting my new partner to all things green (Trevor, you are such a good sport!). Following the advice I always give to others, I have implemented a plan that has me continually greening my life, and I congratulate myself on making changes not all at once but by plugging away at it, bit by bit.

Part of the bit-by-bit plan includes being aware of how I judge others. I try to open conversations about more environmentally responsible options rather than just tell people they are doing it wrong. Let's educate people rather than perpetuate shame.

As I said in the first edition of this book, few of us have the scientific background or the time to research and come to a definitive conclusion about what the "best" environmental services, products and programs are. Still, I believe the most important thing for us to know is that we do have options. The simple choices you make every day – what you buy at the grocery or hardware store, whether or not your drive your car to work, whether you use tap water or bottled – make a difference. I'm not interested in debating about climate change or any large-scale environmental trends in this book. Rather, I will assert this: the average North American's current lifestyle is not sustainable. We are wasteful. The planet does not have unlimited resources. We need to be more careful with the natural resources we have in order to ensure that they are clean, healthy and abundant in the future.

HOW TO USE THIS BOOK

Live Green, Calgary! is organized into chapters that apply to particular aspects of our lives. Each chapter has four sections:

- Get on the Green Team
- Products and services in Calgary
- Resources
- Rebates and government programs

Each product, service, resource or program listed includes contact information. That said, communicating web links in print is a challenge: they are often long and unwieldy, and therefore I have tried to direct you to the right information rather than suggest you type complex site addresses.

To make this easier, a website companion to this book has been created with all of the links available to you, as well as updates as they become available. Go to www.livegreencalgary.com and log in using the password EarthWise. The hyperlinks will take you where you want to go.

WHO IS INCLUDED IN *LIVE GREEN, CALGARY!*

The vital qualification for inclusion in this guide is that the green product, service, resource or program be available and applicable to Calgarians. As I amassed information for this book, I considered a product, service, resource or program to be environmentally friendly if 80 per cent of what it offers includes at least one of the following factors:

- organic;
- locally produced;
- renewable, reusable, recyclable;
- made from reclaimed or sustainably harvested materials;
- helping conserve resources or reduce waste;
- non-toxic; and
- fair trade.

Or, if its

- mission/vision is to help encourage more environmentally friendly lifestyles;
- waste is handled in an environmentally responsible manner; and
- manufacturing/operations are particularly resource efficient.

Or, if it is the best option where no others exist.

When it comes to living greener and following through with it, knowledge about our communities and the resources they provide is key. The purpose of this book is to lay bare the information that is out there in a reader-friendly manner. *Live Green, Calgary!* strives to be a complete listing of environmentally sensitive products, services and programs available to Calgarians that imparts precise, applicable information that will save you time as you begin your quest for a sustainable lifestyle. (And there are money-saving tips, too!)

When you are making everyday decisions – what food to buy, what neighbourhood to live in, how to commute – *Live Green, Calgary!* will help you discover your options. That way, you will be able to do the best you can with the resources you have.

HOME HEATING

GET ON THE GREEN TEAM

Improving your home's building envelope is the most cost-effective way to conserve energy and save money, and should be your first step toward a greener home. Stopping air leaks will reduce drafts and heat loss, protect the building and minimize the amount of noise and dust that enter your home. Consider:

- checking windows and exterior doors for leaks; caulk and weather-strip them as necessary.
- replacing single-pane windows with more efficient ones or adding inexpensive window films or storm windows in the winter.
- replacing inefficient exterior doors with insulated-core doors or adding storm doors.
- keeping window coverings open in winter during the day to let the sun help heat your home; close them when the sun sets to keep the heat in.

You can also heat your home more efficiently by:

- installing an ENERGY STAR® programmable thermostat to lower the temperature at night and when nobody is home; this will help you save at least 10 per cent on heating bills.
- replacing dirty furnace filters, which only make your furnace work harder; replace filters according to your furnace manufacturer's recommendations (usually every one to three months during the heating season).
- using a washable furnace filter instead of disposable ones (both are available from your local building supply store); but keep in line with manufacturer recommendations.
- keeping air vents free of obstruction so warm air circulates easily.

- installing ceiling fans to circulate air and even out temperatures in both winter and summer.
- considering alternative ways to heat your home, especially if you are building a new home. (Passive solar design, geoexchange (geothermal), solar hot water heating … there are lots of options!)
- choosing an ENERGY STAR® high-efficiency furnace model when it comes time to replace your furnace.

Eco-words

Building envelope: what separates the interior from the exterior of a building. It is comprised of the foundation, roof, walls, doors and windows. The integrity of the building envelope is essential to temperature and moisture control within a building.

PRODUCTS AND SERVICES IN CALGARY

1. BLOWER DOOR TESTING

Have a certified energy adviser perform a blower door test, using a powerful fan to identify places in your home where warm air can escape. The following companies are approved by the Government of Canada to perform ecoENERGY evaluations, which include blower door tests:

AmeriSpec of Calgary South
www.amerispec.ca (choose Book an ecoENERGY Evaluation to find the representative nearest you)

ATCO EnergySense
www.atcoenergysense.com (choose Residential Services, then Blower Door Assessments) or 403-310-7283

Ecofinity Canada Inc.
www.ecofinity.ca or 403-699-9700

VerdaTech Inc.
www.energyexperts.ca (choose Residential, then click on the province of Alberta to find Blower Door Testing) or 403-251-0683

2. RENEWABLE ENERGY SYSTEMS

Clean Energy Developments designs, engineers and installs geoexchange heating and cooling systems. Their technicians are certified by the Canadian Geoexchange Coalition.
www.cleanenergydevelopments.com or 403-244-0111

Ener-West Geo-Energy Services designs and installs geoexchange (geothermal) heating and cooling systems and has expanded to implement solar hot-water systems to complement their geoexchange applications.
www.ener-west.com or 403-279-2384

ENMAX wants to help Albertans generate their own heat and electricity using renewable energy sources. This includes solar hot-water systems that can be used for home heating.
www.enmax.com/solar or 310-2010

Greenergy Renewable Energy Solutions designs and implements solar electric, solar thermal, micro-hydro and wind systems with the goal of helping Calgarians reduce their demand for conventional fossil fuel and therefore their ecological footprint.
www.greenergycanada.ca or 403-472-7001

Ground Source Energy provides geoexchange systems that are site-specific to meet your needs, from the fieldwork to the heat pump.
www.groundsourceenergy.com or 403-874-6259

REACT Energy designs and installs geothermal heating and cooling solutions for both residential and commercial projects.
www.reactenergy.ca or 403-873-0109

Simple Solar Heating builds and assembles solar heating systems for domestic space heating and hot-water use. They focus on simple, reliable, economical systems for the average homeowner.
www.simplesolar.ca or 403-938-5803

SkyFire Energy Inc. designs and installs a variety of renewable energy systems for both residential and commercial purposes. They specialize in solar thermal, solar electric and wind power.
www.skyfireenergy.com or 403-251-0668

Solar Services concentrates on solar hot-water heating, as they feel that's where people can get the biggest bang for their alternative-energy buck. They provide solar hot-water heating solutions for residences, agricultural applications, government buildings and swimming pools.
403-542-7657

> **Green Fact**
> The City of Calgary is trying to make it easier for people to install solar panels on their homes by removing the requirement for a development permit to do so. Other permits, such as building, electrical and plumbing, are still required depending on what type of system you are installing. Your panel installer will be able to tell you what you need and guide you through the process.

3. THERMAL IMAGING AND THERMAL ENERGY AUDITS

Calgary Thermal Imaging and Thermal Energy Audits use thermal imaging to identify air leaks and weak insulation in your home. See exactly where your home is losing energy and costing you money.
www.calgarythermalimaging.com or 304-259-9181
www.thermalenergyaudits.ca or 403-200-7082

RESOURCES

1. ATCO ENERGYSENSE

ATCO EnergySense has a wealth of unbiased information on how to reduce energy use. Use their energy scorecard to do your own home energy audit and create an action plan. Or call them for advice on your particular energy efficiency issue.

www.atcoenergysense.com (search "energy scorecard") or 403-310-7283

2. NATURAL RESOURCES CANADA

Natural Resources Canada provides free software that evaluates domestic energy production, savings, costs, emission reductions, financial viability and risk for renewable energy and energy-efficient technologies.

www.retscreen.net

3. OFFICE OF ENERGY EFFICIENCY

The Government of Canada's Office of Energy Efficiency is an amazing source for information. However, its search functionality is very limited. Therefore, websites included here are very cumbersome. Google search tips are included as an alternative to typing in these websites. Visit www.livegreencalgary.com for direct links to these sites.

Performing simple tests for leaks:
www.oee.nrcan.gc.ca/residential/personal/new-home-improvement/choosing-insulation-sealing/air-leakage/khi-findleak.cfm?attr=4
or Google "finding leakage areas oee Canada"

Sealing leaks:
www.oee.nrcan.gc.ca/residential/personal/new-home-improvement/choosing-insulation-sealing/air-leakage/khi-caulking-sealing-materials.cfm?attr=4
or Google "caulking and sealing oee Canada"

Improving attic insulation can reduce energy use by 15 per cent, and up to 30 per cent of your home heating is lost through a non-insulated basement. See:
www.oee.nrcan.gc.ca/residential/personal/new-home-improvement/choosing-insulation-sealing/materials/khi-intro.cfm?attr=4
or Google "choosing insulation sealing attic oee Canada" or "choosing insulation sealing basement oee Canada"

Making windows and doors more efficient:
www.oee.nrcan.gc.ca (under Personal Use click Residential, then Windows, Doors and Skylights)

> **Green Fact**
> Heating accounts for an average of 60 per cent of the energy used in your home.

REBATES AND GOVERNMENT PROGRAMS

1. ALBERTA ENERGY-EFFICIENCY REBATES

Receive rebates from the government of Alberta to help make your home more energy efficient. Buy a qualifying energy-efficient furnace or hot-water heater, add insulation and get a home energy audit and you can receive up to $4,250!
www.climatechangecentral.com (under My Rebates)

2. ECOENERGY RETROFIT PROGRAM

Homeowners can receive federal grants up to $5,000 from the Government of Canada's ecoENERGY Retrofit Program when they improve the energy efficiency of their homes.
www.oee.nrcan.gc.ca (click ecoENERGY Retrofit)

INDOOR WATER USE

GET ON THE GREEN TEAM

Water is precious, and there are many simple ways to reduce your water use. You can make a difference!

First, fix the leaks. A faucet losing one drop per second wastes up to 25 litres of water per day, or over 10,000 litres per year!

Next, consider the amount of water you use and how you can cut down. Taking a shorter shower under a low-flow showerhead can make a big difference to the amount of water you use. The average bath uses 100 litres of water, while low-flow showerheads use 7.5–10 litres per minute. A five-minute shower saves half the water of the average bath. You can also install aerators on bathroom faucets to reduce flow by 25 to 50 per cent and you won't even notice the difference!

After faucets, consider your toilets. Check for toilet leaks by adding a squirt of food colouring to the toilet tank. Wait 10 minutes and look in the bowl. If the water in the bowl is red or green or blue (or whatever food colour you used), you have a leaky toilet.

Instead of fixing that leaky, 13+-litre-per-flush toilet, you might want to replace it with a low-flow, six-litre-per-flush model which will use at least 50 per cent less water. Better yet, think about installing a dual-flush toilet, which will use only three or four litres for a liquid flush and six litres for a solid.

If you're not quite ready to spend money on a new toilet, you can make inexpensive modifications to your existing toilet so it uses less water. For example, toilet dams are inexpensive and save about five litres of water per flush. These are barriers placed inside the toilet tank to displace water, making the volume of the tank smaller so that it uses less water. You can also displace water in the toilet tank by inserting bottles or containers. How much water you save depends on how big the displacement device is. Never use a brick as a displacement device, as it can harm your plumbing. You can also install alternative flushing devices which close the flush valve after the tank is only partially emptied. For example, a dual-flush modification device lets the flusher choose a light or a full flush. Remember: if you have to double flush your toilet, whatever you've done to modify it is not working properly and is in fact defeating the purpose of your conservation efforts. Be sure to monitor any changes you've made and adjust as necessary.

Besides fixing faucets and toilets, you may wish to consider the following water-conscious tips to ensure you are cutting back on your use of H_2O.

- Get rid of your garburator. It uses a lot of water and electricity to liquefy organic scraps and increases the waste-water treatment required. Compost instead!
- Be aware of running water.
- If you let the water run while brushing your teeth or shaving you can let 45 litres of water run down the drain in five minutes.
- Keep drinking water in the fridge so you don't have to run the tap to get cold water. Water plants with unused drinking water.
- Wash vegetables in a shallow sink of water rather than under running water. Better yet, wash them in a bowl of water and when finished use that water on house or garden plants.
- Know how much water you use, but also be aware of what you put down the drain. Use non-toxic cleaning products for your body, laundry, dishes and household cleaners.
- Use less hot water.
- Wash clothes in cold water.
- Check the setting for your hot-water tank: 49–54 degrees Celsius is hot enough. Experiment based on your needs.

- Insulate the pipes from your hot-water tank to help reduce the delay in getting hot water to your taps. Less time running water while waiting for it to heat up means less water wasted.

> **$ave Green**
> The sanitary and storm sewer charges on your ENMAX bill are both based on a charge per cubic metre of water that comes *into* your home, not what actually goes down the drain. So if you conserve water, you will save money on both the clean water coming into your home and the waste water leaving it.

PRODUCTS AND SERVICES IN CALGARY

1. COMPOSTING TOILETS

A composting toilet is perfect for areas where you can't connect to a sewer or septic system, or in areas of water shortage. To see one in action, tour the Alberta Sustainable Home/Office, at **www.ecobuildings.net**. Get one from Let's Go Green or Sunergy Systems, which has been designing resource efficient buildings since 1976:
www.letsgogreen.com or 1-888-248-9754
sunergy@compostingtoilet.com or 403-637-3973

> **Eco-words**
> *Composting toilet:* a toilet that uses air and bacteria rather than water to break down toilet waste.

2. PLUMBING

The plumbing department of your local building supply store and Clean Calgary's EcoStore carry toilet modification devices that will reduce the water your 13+-litre-per-flush toilet uses.
www.cleancalgary.org or 403-230-1443

> **Green Fact**
> Canadians are the second largest per capita users of water in the world. Calgarians use about 340 litres of water per person per day in their homes. In the UK, each person uses about 147 litres per day.

RESOURCES

1. "LEAKY" WORKSHOPS AND TIPS

The City of Calgary can help you to fix that leaky faucet or toilet. Attend a leak workshop or find repair tips online.
www.calgary.ca/waterworks (search "leak workshop")
www.calgary.ca/waterworks (search "water leaks and repairs")

2. LOW-FLOW INFO

Find the best low-flow and dual-flush toilets on the market in the free "Maximum Performance: Testing of Popular Toilet Models" report.
www.cwwa.ca (click on CWWA Reports on 6 Litre Toilets and Drainlines)

3. ON-DEMAND HEATING

Consider making your next hot-water heating system an on-demand system and eliminate the cost of keeping water hot when it's not being used. Learn about the advantages and disadvantages at:
www.atcoenergysense.com (under Tools and Resources choose Articles and Fact Sheets, then Residential and look under Hot-water Heating)

> **Green Fact**
> In Calgary, all new homes and commercial buildings, as well as renovations that require a plumbing permit, must install water-conserving fixtures such as low-flow toilets.

REBATES AND GOVERNMENT PROGRAMS

1. MUNICIPAL TOILET REBATE PROGRAMS

Replace your water-guzzling toilet in the following municipalities and receive a rebate:

City of Airdrie: $50 to $100 rebate
www.airdrie.com/environmental_services (click on Residential Toilet Replacement Program Details)

The City of Calgary: $50 rebate
www.calgary.ca/waterservices (click on Residential Toilet Replacement Program)

Town of Cochrane: $100 to $250 rebate
www.cochrane.ca (search "toilet rebate program")

The Town of Okotoks: $50 rebate
www.okotoks.ca (search "toilet rebate")

> **$ave Green**
> Buy a qualifying ENERGY STAR® clothes washer and receive a $100 rebate. **www.climatechangecentral.com** (under My Rebates)

ELECTRICITY

GET ON THE GREEN TEAM

Most of the electricity in Alberta comes from coal-fired generation, which is extremely polluting. You can cut back on dirty electricity by supporting renewable electricity initiatives, such as wind and solar power, which are clean and renewable sources of electricity.

Regardless of the electrical energy source you use, it is important to cut back on the amount of electricity used in your household. Eliminate phantom loads by using a power bar with an on/off switch to ensure your electronics are actually turned off and not just in standby mode. This can make a real difference, especially in areas like your computer workstation and entertainment system. When not in use, turn these devices off at the power bar.

Use a Kill-a-Watt to determine which devices in your home carry a phantom load and how much electricity they use. Check everything from your cordless phone to your video gaming system to the refrigerator and see how much power things use even when turned off! Sign a Kill-a-Watt out free from the Calgary Public Library.

In addition to eliminating phantom loads, you can cut back on electrical energy consumption by seeing the light! Use energy efficient light bulbs, such as those rated as ENERGY STAR®, instead of incandescent bulbs. Compact fluorescent (CFL), high-efficiency halogen or light-emitting-diode (LED) lights all use less energy than conventional incandescent bulbs. Incandescent light bulbs convert only 5 per cent of the energy they use into light, whereas LEDs are up to 90 per cent efficient.

Use motion and occupancy sensors, timers, dimmer switches and the lowest-watt bulb, to match your lighting to your needs. Lampshades with bright liners maximize reflection to make the most of the light being produced. Muted colours on walls and ceilings reflect much more light than dark colours, making rooms brighter.

> **Eco-words**
> *Light Emitting Diode (LED):* when electricity passes through certain types of semiconductor devices, energy is released in the form of light. LEDs are up to 90 per cent efficient, don't have a filament that will burn out, don't get very hot and are very durable. LEDs are commonly used for traffic lights and emergency exit signs.

PRODUCTS AND SERVICES IN CALGARY

1. ECOLOGO™ ENERGY PROVIDERS

EcoLogo™ helps consumers make more environmentally conscious decisions by providing third-party certification of the world's most sustainable products. The following companies provide EcoLogo power.

Bullfrog Power provides EcoLogo-certified wind power generated right here in Alberta. When you sign up with Bullfrog, the company feeds renewable, emission-free wind power into the provincial electricity grid on your behalf.
www.bullfrogpower.com or 1-877-360-3464

> **$ave Green**
> Save $10 on your first bill and support the Pembina Institute when you purchase renewable wind power for your home from Bullfrog Power. Enter the promo code "PEMBINA" when signing up through www.bullfrogpower.com.

Canadian Hydro generates EcoLogo-certified electricity in its wind, hydroelectric and biomass facilities. Purchase renewable energy

certificates that represent clean energy added to the grid equal to the amount coal-generated electricity used in your home.
www.canhydro.com or 403-269-9379

> **Green Fact**
> In April 2007, the Government of Canada announced it will set national performance standards to phase out inefficient light bulbs, such as incandescents, by 2012.

When you purchase gas and electricity together under ENMAX's EasyMax program, you are guaranteed to save $100 a year. Choose to apply those savings to Greenmax and 5000 kWh of EcoLogo-certified renewable energy will be purchased on your behalf. This equals roughly two-thirds of the electricity used by the average household annually.
www.enmax.com/gogreen (click Greenmax) or 310-2010

Seventh Generation Power purchases 100 per cent renewable energy from local sources to help the average person or business make electricity choices that will positively impact future generations.
www.7gpower.com or 403-668-4979

2. OTHER RENEWABLE ENERGY SOURCES

> **Green Fact**
> In 2007, ENMAX started providing The City of Calgary with enough wind power to meet 75 per cent of its corporate electricity needs. The City's goal is to increase the amount of electricity it uses from green sources to 90 per cent by 2012. This will help meet its goal of reducing its corporate greenhouse gas emissions to 50 per cent below 1990 levels by 2012.

ENMAX wants to help Albertans generate their own heat and electricity using renewable energy sources. This includes solar hot water systems that can be used for home heating.
www.enmax.com/solar or 310-2010

At Goose Creek Renewable Energy they practise what they preach by living off the grid themselves! Contact them to create a system that allows you to get electricity from renewable sources such as the sun and wind.
www.goosecreek.ca or 403-684-3730

SkyFire Energy Inc. designs and installs a variety of renewable energy systems for residential and commercial purposes. They specialize in solar thermal, solar electric and wind power.
www.skyfireenergy.com or 403-251-0668

SolarVantage is passionate about bringing efficient, reliable, renewable power to residential, commercial and agricultural properties. Look to them for renewable energy products, consulting, installation and maintenance services.
www.solarvantage.ca or 403-874-0657

3. KILL-A-WATT

Borrow a Kill-a-Watt for free from the Calgary Public Library and see how much power your electronic devices use, even when they're turned off. Just plug the Kill-a-Watt into the electrical outlet, and then plug the device into the Kill-a-Watt. Take a reading of how much electricity the device is using when it's on, then turn it off and see how much it uses when you've turned it off. If it is using electricity when turned off, plug it into a power bar that has an on/off switch, plug that into the Kill-a-Watt and then turn the switch off. You will see that the electricity use is now zero.
www.calgarypubliclibrary.com (search "kill a watt")

> **Eco-words**
>
> *Phantom load:* the power consumed by any electronic device while it is switched off. Often, electronics such as remote controls, LED displays and automatic timers continue to draw power even when they are "off."

RESOURCES

1. ENERGY STAR®

ENERGY STAR® identifies the most energy-efficient products, from appliances to electronics and windows, and from front doors to entire homes.
www.oee.nrcan.gc.ca/residential/energystar-portal.cfm

2. FEDERAL GOVERNMENT RESOURCES

The Government of Canada's Office of Energy Efficiency website explains everything you need to know about choosing energy-efficient lighting for both indoors and out.
www.oee.nrcan.gc.ca (under For Personal Use click Residential, then Lighting)

Natural Resources Canada provides free software that evaluates the energy production, savings, costs, emission reductions, financial viability and risk for renewable energy and energy-efficient technologies.
www.retscreen.net

3. THE LIGHT BULB CALCULATOR

Use ENMAX's light bulb calculator to find out how much money and electricity you can save by switching to compact fluorescent light bulbs.
www.enmax.com/gogreen (choose Conservation)

Green Fact

The City of Calgary recommends leaving compact fluorescent light bulbs (CFLs) for pick up with your regular household garbage, treating them as you would broken glass or wrapping them completely in duct tape. CFLs can be recycled at Ikea and Home Depot locations.

Green Fact

If you break a CFL it needs to be disposed of carefully. The Government of Canada's Office of Energy Efficiency recommends:

Sweep – don't vacuum – all of the glass fragments and phosphor powder. Place everything you swept up in a plastic bag.
Wipe the area with a damp paper towel to pick up stray shards of glass and fine particles.
Put the used towel in the plastic bag.

In accordance with recommendations from The City of Calgary, dispose of the remnants in your household garbage, treating it carefully as you would other broken glass.

Electricity

HOUSEHOLD CLEANING

GET ON THE GREEN TEAM

Household cleaning products are some of the most significant influences on indoor air quality. Using non-toxic cleaners is one of the easiest and most effective ways to keep indoor air safe for your family. Furthermore, many conventional cleaning supplies have hazardous chemical warnings on them: corrosive, flammable, poison, explosive. There are safer alternatives for cleaners and deodorizers.

The "clean smell" of conventional cleaning products can be toxic. Most often the smell comes from synthetic fragrances, which can affect the central nervous and respiratory systems and potentially cause cancer. This is true for air fresheners as well as cleaning products.

You can also improve air quality in your home by investing in a good vacuum cleaner. Choose a central vacuum system and ensure it is properly vented to the outside, or choose a sealed HEPA system vacuum, which is completely sealed to ensure that all the air that is sucked in goes through the filter and doesn't leak back out. Both these options ensure that all the particles you have vacuumed up are not immediately released into the surrounding air.

PRODUCTS AND SERVICES IN CALGARY

1. CLEANING PRODUCTS

Calgary's own Claudia's Choices has their own line of "envirosponsible" laundry detergent, free of phosphates, dyes, bleach and animal by-products. It is made in Canada and the company gives a discount to those who bring back the bucket packaging for reuse. They also

carry eco-friendly cleaning and baby products. Find retailers or order online at:
www.claudiaschoices.ca or 1-888-613-9274

Small Planet specializes in cleaning products that are effective and free of synthetic ingredients. They sell tile and tub cleaners, all-purpose cleaners, carpet deodorizers, furniture polish and more, all made in Calgary. Order online or find the retailer nearest you at:
www.smallplanet.ca or 403-863-7672

> **Green Fact**
> In Canada, manufacturers of household cleaners are not obliged to list the ingredients in their products.

2. CLEANING SERVICES

No Cleaner Maid Service provides residential cleaning services without the use of chemical cleaners that will harm the Earth.
www.nocleanermaidservice.ca or 403-452-8640

Skylark Cleaning Services provides residential and commercial cleaning services using only environmentally friendly cleaning products that protect both the environment and our health.
403-978-1611

> **Green Fact**
> Most conventional household cleaners are made from petroleum products.

RESOURCES

1. GUIDES

The following online guides to green cleaning are the best I've found:

Clean Calgary Association's *Green Cleaning Guide* helps you know what to look for and what to avoid in cleaning products, as well as simple recipes for homemade green cleaners.
www.cleancalgary.org click Guides

Environmental Defence's *Toxic Nation Guide to Spring Cleaning* teaches you about the health risks associated with ingredients found in many conventional household cleaners, and introduces some healthier alternatives.
www.environmentaldefence.ca (search "spring cleaning")

Guide to Less Toxic Products, produced by the Environmental Health Association of Nova Scotia, provides lots of information about the risks of and alternatives to a wide variety of common household cleaners.
www.lesstoxicguide.ca (choose Household Cleaners)

2. WORKSHOPS

Clean Calgary's green cleaning workshop demonstrates how using greener cleaners improves indoor air quality. Learn how to evaluate cleaning products for their greenness, and learn how to make your own green cleaning products.
www.cleancalgary.org (click Education, then Workshops and Courses for Individuals) or 403-230-1443

REBATES AND GOVERNMENT PROGRAMS

1. CHEMICAL DISPOSAL

Take toxic household chemicals, including cleaners you don't want to use, to one of The City of Calgary's eight year-round drop-off depots for free, safe disposal. See Appendix.
www.calgary.ca/waste (choose Household Chemical Drop-off Program or call 3-1-1)

YARD AND GARDEN

GET ON THE GREEN TEAM

How much lawn do you need? A perfect lawn requires a lot of resources, including water, fertilizer and your time and energy to care for it. Think about how you use your yard, and replace that Kentucky bluegrass that isn't being used for play or social activities with low-water-use native plants and groundcovers, or perhaps a vegetable garden. Native plants are naturally hardy to Calgary's specific climatic zone, which means they do not require water above and beyond that which falls from the sky.

While you are thinking about rainfall, consider buying rain barrels. Rain barrels catch water flowing from your eavestroughs, which you can then use to water your yard and garden. Plants love the soft, non-chlorinated water, and you will love saving on your water bill!

Even if you do water your plants from the rain barrel, always water efficiently. Avoid watering in the heat of the day, put a timer on the sprinkler and, most important, find out how much water your plants really need before pouring it on.

Your lawn and garden will need less water if you have healthy soil. Healthy soil is the most important component of a healthy yard. Non-toxic fertilizers are available: liquid seaweed, manure and compost all add valuable nutrients to your soil.

While you are considering non-toxic fertilizers, think about using non-toxic methods to discourage common pests in your yard, too. Remember that each one of these pests is part of the eco-system. The despised wasp, for example, helps maintain a balanced population of aphids. Ask a garden expert how to use everyday kitchen ingredients like soap, baking soda and onions as natural "pest" inhibitors. For example, a saucer of beer will trap slugs!

Finally, not many people consider how landscaping can help you save on your home heating and cooling costs. Deciduous trees, such as Shubert chokecherry and columnar aspen, shade windows in the summer while letting sun shine through in the winter. Evergreen trees, such as like Colorado spruce or Scots pine, act well as wind blocks.

> **Green Fact**
> Using a power lawn mower for one hour releases the same amount of pollution as driving a car 563 kilometres. Gas powered leaf blowers and edging machines are also sources of air and noise pollution.

PRODUCTS AND SERVICES IN CALGARY

1. COMPOST BINS AND RAIN BARRELS

Clean Calgary's The EcoStore has rain barrels for sale. They also have composters, subsidized by The City of Calgary. Compost is great for your garden and helps you reduce waste headed for the landfill. You can also have a composter shipped to your doorstep by purchasing it through The City of Calgary Online Store.
www.calgaryonlinestore.com (search "composter").
www.cleancalgary.org (click The EcoStore) or 403-230-1443

2. ECO-FRIENDLY LANDSCAPING SERVICES

Eco-yards designs and installs environmentally friendly, low-maintenance landscapes. They use locally sourced stone, locally created compost and their own Eco-Yards Spray to create outdoor spaces that are safe and healthy for everyone.
www.eco-yards.com or 403-969-1176

EnviroPerfect Solutions provides organic lawn care services. They develop great-looking lawns by focusing on restoring the natural biology rather than using toxic chemicals.
www.enviroperfectsolutions.com or 780- 447-9600, 1-877-377-4769

Heron's Nest Landscape Design specializes in eco-urban landscape design and sustainable landscapes.
403-347-4155

3. ECOROOF INITIATIVE

The Alberta EcoRoof Initiative educates Calgarians about green roof technologies and how they work here in Calgary as part of a green building strategy.
www.calgarytechnologies.com (under Office & Meeting Facilities click Green Initiatives)

4. GARDEN CENTRES

Bowpoint Nursery specializes in cultivating and selling the native woody plants of southern Alberta. They also sell compost, soil amendments and mulch created from yard waste collected by local landscapers.
www.bowpointnursery.com or 403-686-4434

Greengate Garden Centres is one of the only garden centres in the country that accepts plastic plant trays and pots back from its customers for recycling. They also encourage people to use alternatives to conventional pesticides by selling ladybugs, corn gluten weed and feed, and organic fertilizers.
www.greengate.ca or 403-256-1212

5. GOOD STUFF FOR YOUR GARDEN

If you haven't started composting yet but want the good stuff for your garden, you can buy natural, locally produced soil amendments instead of conventional fertilizers from companies like The Worm Croft, Western Canada Compost and Worms@Work.

www.thewormcroft.com or 403-461-2782
www.westerncanadacompost.com or 403-251-9639
www.wormsatwork.com or 403-730-3405

6. MUCHO MULCH

Free mulch – derived from Christmas trees collected by The City of Calgary – is available from the East Calgary Landfill while quantities last. Calgarians are welcome to take one or two truckloads for residential use.
17 Avenue and 68th Street SE or 3-1-1

ECCO Chips is a wood mulch product made in Calgary from post-consumer waste wood. It comes in five colours, using pigment from minerals and vegetables.
www.eccochips.com or call 403-720-0442

> **Eco-words**
> *Mulch:* the application of organic materials like bark, wood chips or stones placed over soil to reduce water evaporation, inhibit weed growth and minimize erosion.

7. NATIVE PLANTS AND SEEDS

ALCLA Native Plant Restoration specializes in growing and harvesting native wildflowers, shrubs and grasses. They also provide consulting services on using native plants in home landscaping and green roofs, as well as in commercial, public and industrial settings.
www.alclanativeplants.com or 403-282-6516

Seedy Saturday is about saving, swapping, buying and selling seeds that are open-pollinated and have not been genetically modified. Watch for it every spring.
www.calhort.org (under Gardening Resources click on Community Gardens)

8. STARBUCKS GROUNDS

Take advantage of Starbucks' Grounds for Your Garden program. They will give you the grounds from their famous coffee for free! Coffee grounds make a nutrient-rich mulch in the vegetable garden and can be used as a "green" in your compost bin. Call the manager of the store nearest you to see if they participate in the program.
www.starbucks.ca

9. WATER SAVINGS

Buy an Outdoor Water Saver Kit from The City of Calgary. The kit includes a precision garden hose nozzle, an automatic sprinkler timer and a rain gauge.
www.calgary.ca (search outdoor water saver kit)

RESOURCES

1. GARDENING TIPS

Canada Mortgage & Housing Corporation has lots of free information on how to make your yard more environmentally friendly. Find out how to create a rain garden to improve storm-water management, how to create a low-maintenance lawn, and more.
www.cmhc.ca (click Consumers, then Maintaining a Home, then Landscaping)

Coalition for a Healthy Calgary has information, ideas and resources to help you keep a pesticide-free yard.
www.healthycalgary.ca (click How to Go Natural)

Evergreen is a non-profit organization that provides practical tools for naturalized gardening through its Home Grounds program. Check out their Native Plant Database, along with many natural gardening tips and tricks.
www.evergreen.ca

2. COMPOSTING

Got a composting question? Clean Calgary Association is happy to help you with your composting questions and problems. Call their Composting Hotline, take a composting workshop or check out their *Compost Guide*.
www.cleancalgary.org (click Guides) or 403-230-1443

If you're not composting but want to keep your yard waste out of the landfill, take it to the East Calgary Landfill. At the scale house, tell them you have organic yard waste you want to compost and they will direct you to the right area.
17 Avenue and 68th Street SE or 3-1-1

> **Green Fact**
> Non-synthetic wine corks can be composted. Tear them up and toss them into your backyard heap.

3. WATER-SAVVY

The City of Calgary's Water Wise Gardening brochure will teach you water-savvy habits specific to your local green space.
www.calgary.ca/waterservices (click Water Conservation, then Lawn & Garden and look for the download on the lower right-hand side of the page, or search "water wise gardening")

REBATES AND GOVERNMENT PROGRAMS

1. HEALTHY YARDS

Join The City of Calgary's Healthy Yards Program and learn all you can about environmentally friendly yard and garden care in Calgary. Receive free training on how to compost, conserve water, manage pests and grasscycle. And get a free rain barrel and composter!
www.calgary.ca (search "healthy yards program")

> **Eco-words**
> *Grasscycle:* the practice of recycling grass clippings by leaving them on the lawn. It acts as a slow release fertilizer, limits moisture evaporation from the soil and eliminates bagging and raking (not to mention reducing the amount of yard waste entering landfills).

2. RECYCLING YARD WASTE

Recycle your Christmas tree, bagged leaves and sagging pumpkins free of charge at one of The City of Calgary's many seasonal drop-off locations or at any of the three landfills. The East Calgary Landfill accepts leaves and pumpkins without charge year round.
www.calgary.ca (search "leaf and pumpkin")

3. SAVE ON TREES

The City of Calgary's Planting Incentive Program encourages planting trees on residential City property by subsidizing (on approved application) 50 per cent of the cost to purchase and plant certain trees. Applications must be made by June 30 each year.
www.calgary.ca (search "planting incentive program") or 3-1-1

HOME BUILDING AND RENOVATION

GET ON THE GREEN TEAM

The single biggest energy users in our personal lives are our homes. Therefore, building or renovating your home provides the biggest area of opportunity to live a more environmentally responsible lifestyle. Check out these examples of energy-efficient housing in the Calgary area for inspiration when building or renovating your home:

The Alberta Solar Decathlon team competed to design, build and operate the most attractive, practical and energy-efficient solar-powered home in the 2009 Solar Decathlon challenge. Check out their uniquely Albertan design.
www.albertasolardecathlon.ca

The Calgary EcoHome is located in the northwest's Scenic Acres and has no gas, water or sewer lines. Energy conservation, rainwater capture and renewable energy sources are the keys to operating this self-sufficient home. Take one of their frequent tours to see the composting toilet and other sustainable features in action.
www.ecobuildings.net or 403-239-1882

Drake Landing Solar Community in Okotoks was designed to meet 90 per cent of its residential space heating needs by solar thermal energy.
www.dlsc.ca

Echo Haven is a community of 25 healthy homes whose goal is to reduce average use of electricity by 80 per cent and water by 72 per cent. It is also a demonstration site for CMHC's EQuilibrium initiative,

where homes produce as much energy as they consume and can feed excess power back into the grid.
www.echohaven.ca
www.cmhc-schl.gc.ca (search "EQuilibrium")

The above examples are the vanguard of eco-housing. Incorporate some, or all, of the elements of these houses when building or renovating. Also, remember to choose wood products – furniture, building supplies etc. – that are sustainably harvested. Look for certification from a group like the Forest Stewardship Council (FSC), an international organization that sets standards for responsible forest management. The FSC label enables consumers worldwide to recognize products that support those standards.

PRODUCTS AND SERVICES IN CALGARY

1. ECOROOF INITIATIVE

The Alberta EcoRoof Initiative educates Calgarians about green roof technologies and how they work here in Calgary as part of a green building strategy.
www.calgarytechnologies.com (under Office & Meeting Facilities click Green Initiatives)

> **Eco-words**
> *Green roof:* a building roof that is covered with soil and vegetation planted over a waterproof membrane. Green roofs can help manage and treat storm water, as well as filter pollution.

2. HOME DESIGN AND BUILDING

AlternaHome Solutions was chosen to build a CMHC NetZero demonstration home in New Brunswick and has brought that experience west. They firmly believe that the first step in sustainable construc-

tion is to evaluate the chosen site in order to incorporate solar, wind and geothermal advantages into the building.
www.alternahome.com or 403-461-7188

Avalon Master Builder has a vision of building all of their homes as Net Zero homes at no additional cost to the consumer by 2015. That means all their homes will create as much energy as they use. Every home they build is certified Built Green™.
www.avalonmasterbuilder.com or 403- 226-3485

Bow Crow Design specializes in designing buildings that step away from conventional sources of energy. They designed the CMHC EQuilibrium home at Echo Haven, whose energy-efficient design, combined with passive and active solar heating, will produce as much energy as it consumes.
403-638-3737

Cambio Contracting provides general contracting and project management services. They specialize in strong building structures, proper thermal envelopes and efficient heating systems.
www.cambiocontracting.com or 403-899-2295

Designworks Architecture has Leadership in Energy & Environmental Design (LEED)-accredited architects who take into account everything from the features of the building site to the needs of the end users to create the most energy-efficient building designs possible.
403-282-6776

Ecological Homes creates affordable, ready-to-assemble home packages that are designed to be very efficient. They are designed to achieve an EnerGuide rating of at least 80 and accommodate the installation of solar panels.
www.ecologicalhomes.ca or 780-919-5546

Enviro Custom Homes designs and builds houses with a focus on energy efficiency and renewable energy. Moreover, they aim to build homes that are healthy for the body, the spirit and the planet.
403-375-0092

Folio Homes designs and builds homes in inner-city Calgary with a focus on energy efficiency, indoor environment quality and water conservation. At the same time, they emphasize comfort, control and convenience in the home.
www.foliohomes.com or 403-265-5531

Housebrand helps homeowners create more livable, sustainable homes. They evaluate the location of your home and its design, manage construction or renovation, create the interior design and supply furniture.
www.housebrand.ca or 403-229-4330

Kanas designs and constructs environmentally responsible, energy-efficient buildings, from single-family homes to multi-family residences to commercial buildings.
www.kanas.ca or 403-283-2566

Navona believes that green building addresses the structure, the operating systems and the finishing products, and they construct homes taking all these aspects into account. The result is healthy homes that cost less to operate and maintain.
www.Ready2Build.com or 403-995-3388

Studio T Design Ltd. focuses on building high-quality building envelopes in a manner that takes advantage of the microclimate of the building site. The result is a home that reaps the rewards of passive solar heating and daylighting.
403-220-0542

Sunergy Systems has been designing resource-efficient buildings since 1976. They start with a tight building envelope, then add as

many environmentally responsible features as the site and your budget will allow, including solar systems and composting toilets.
sunergy@wildroseinternet.ca or 403-637-3973

Totally Green Alberta is creating a 1000-home development that is affordable and environmentally responsible. Features include wind and solar energy, water recycling, compost processing and much more.
www.totallygreenalberta.com

3. ICFS

Insulated Concrete Forms (ICFs) are becoming increasingly popular as the framework for new homes because they are fire, sound and moisture resistant. Layers of concrete and insulation make ICFs very strong and energy efficient. Calgary area companies specializing in ICFs include ABS Concrete Systems, ECO-Block, Stackit Wall Systems and Superform Products.
www.absconcrete.com or 403-297-9898
www.eco-block.com or 403-616-7845
www.theaesgroup.ca or 403-923-2487
www.superformproducts.com or 403-627-3555

4. INSULATION

Can-Cell Industries manufactures EcoLogo-certified insulation made from recycled paper.
www.can-cell.com or 403.275.4133

EnviroFoam is spray insulation made from soybean oil and recycled plastic bottles. It emits no volatile organic compounds (VOCs) that can irritate respiratory systems, and acts as a vapour barrier as well as an insulator.
www.whyfoamisbetter.com or 403- 668-4886

Plasti-Fab manufactures moulded expanded polystyrene (EPS) insulation that is Canadian made and EcoLogo certified. Suitable for home renovations or new home building.
www.plastifab.com or 1-888-446-5377

5. LIGHTING

Hi-Lite Creations offers a wide variety of LED lighting solutions for home and office applications.
www.hi-litecreations.com or 403-242-9522

Nemalux is a Calgary-based LED manufacturer that can create an energy-efficient, low-maintenance lighting plan that is right for you.
www.nemalux.com or 403-242-7475

6. PAINT, FLOORING ETC.

ABS Concrete Systems supplies concrete flooring that acts as an excellent heat sink and is low maintenance and extremely durable. Concrete is also often made from recycled material and is recyclable.
www.absconcrete.com or 403-297-9898

Riva's The Eco Store carries sustainable building products such as paint, flooring, insulation and countertops alongside environmentally responsible furnishings and mattresses.
www.rivasecostore.com or 403-452-1001

Spruce up your home with ecocoat 100 per cent recycled latex paint, made in Calgary and available in 14 colours. Get it from Clean Calgary's The EcoStore or direct from the manufacturer.
www.cleancalgary.org or 403-230-1443
www.recyclepaint.com or 403-287-7726

7. ROOFING

GEM Inc. manufactures roof tiles made from recycled tires and plastic that can in turn be recycled at the end of their lives. The tiles are manufactured in Calgary and boast a 50-year warranty.
www.euroslate.ca or 403-215-3333

8. USED BUILDING MATERIALS

Consider used building materials such as windows and doors, plumbing fixtures, kitchen cabinets and more from the Habitat for Humanity Store or Tim's Reusables.

www.calgaryhabitat.ca (click Habitat ReStore) or 403-291-6764

www.timsreusables.com or 403-276-1616

RESOURCES

1. CHECKLISTS, FACT SHEETS AND GUIDES

Canada Mortgage & Housing Corporation has fact sheets describing energy-saving renovation options for many specific styles and ages of homes.

www.cmhc-schl.gc.ca (select Consumers then Renovating a Home, and click Renovating for Energy Savings)

Clean Calgary Association has a couple of guides that will help you make healthy and environmentally responsible choices when building, renovating or renting a home: *Your New Home Guide* and *Green Building and Renovation Guide*.

www.cleancalgary.org (under Guides) or 403-230-1443

Green Alberta is an online resource of green building products and materials that have been independently reviewed to assess their green attributes.

www.greenalberta.ca (under Database choose View the Database)

Reduce the toxic load in your home room by room by going through Toxic Nation's checklist.

www.toxicnation.ca (under Go Toxic Free select Toxic Nation Guides, then Toxic Nation – Make Your Home a Healthy Home!)

2. CONNECT AND DISCUSS

Connect with green neighbours across Canada to learn about green home improvement. Find discussion forums, tools, green building products and more.
www.raisingspaces.com

3. ENERGY-EFFICIENCY RATING AND CERTIFICATION SYSTEMS

Built Green™ is an industry-driven program that provides options for Albertans who want to reduce the impact their new home has on the environment. They certify new homes as Platinum, Gold, Silver or Bronze based on energy efficiency and a checklist of green features.
www.builtgreencanada.ca or 403-808-4258

The Canada Green Building Council Alberta Chapter seeks to transform the built environment so it is environmentally responsible, profitable and healthy. The built community includes homes, commercial and industrial buildings, and entire neighbourhoods. They support programs like the LEED Canada Initiative and the Green Building Performance Initiative, through which builders can certify their buildings.
www.cagbc.org/chapters/alberta

EnerGuide for New Homes is an energy-efficiency rating system for your home. An EnerGuide energy adviser analyzes your new home plans and recommends energy saving changes. After construction, the adviser performs a blower door test to provide a standard measure of your home's energy efficiency.
www.oee.nrcan.gc.ca (under For Personal Use click Residential, then New homes)

EnerVision is a not-for-profit organization whose mission is to make the building community greener. They deliver fine programs such as Building Canada, EnerGuide Rating Service, Built Green™, R-2000 and LEED for Homes.
www.enervisionalberta.com

LEED (Leadership in Energy and Environmental Design) for Homes is a rating system for green building and development practices for single family homes and multifamily buildings up to three storeys.
www.cagbc.org (click LEED Canada, LEED Canada Rating Systems, then Homes)

R-2000 is a national program that promotes the use of cost-effective energy-efficient building practices and technologies. R-2000 outlines how a house must perform rather than specifying exactly how it must be constructed. Typically, a R-2000 home uses 30 per cent less energy than a comparable non-R-2000 home.
www.oee.nrcan.gc.ca (under Personal Use click Residential then ENERGY STAR®, EnerGuide and R-2000)

4. SLOW HOME

The Slow Home movement is built on the idea that suburban sprawl is like fast food – standardized, homogeneous and wasteful – and that it contributes to a lifestyle that is bad for us and the environment. Take the Slow Home Test and learn how your home ranks on the Slow Home principles of "close, simple and light."
www.theslowhome.com

REBATES AND GOVERNMENT PROGRAMS

1. CMHC AND GENWORTH REFUND

Canada Mortgage & Housing Corporation and Genworth Financial Canada both offer a 10 per cent refund on mortgage loan insurance premiums when a borrower buys or builds an energy-efficient home or makes energy-saving renovations to an existing home.
www.cmhc-schl.gc.ca (click Consumers, then Mortgage Loan Insurance and follow the link titled Financing an Energy-Efficient Home or Energy-Saving Improvements is Easier than You think)
www.genworth.ca (search "Energy Efficient Housing Program")

HOME DÉCOR

GET ON THE GREEN TEAM

How you finish and decorate your home can also have a huge impact on indoor air quality. Certain types of glues and finishes release volatile organic compounds (VOCs) that are toxic to the nervous system. If you know "new car" or "new house" smell, you know what VOCs can smell like. But you can make choices in your home that will reduce your exposure to VOCs.

Indoor plants, just like their outdoor cousins, release oxygen into the air as a by-product of photosynthesis. One medium-sized plant per 100 square feet of home or office space absorbs and filters common indoor air pollutants like formaldehyde, benzene and carbon monoxide emitted from building materials, carpets and furniture.

However, to save your plants from having to filter these contaminants, look for paint, floor coverings and furnishings that are low in VOC emissions. More and more low-VOC finishes are becoming available all the time, so be sure to ask your retailer specifically about them.

When choosing paint, you may also wish to consider colour. Light-coloured surfaces reflect more light than dark-coloured ones do. Paint your walls and ceilings in muted tones, using dark colours for accents. This will make maximum use of the light you have and decrease the need for artificial lighting.

Finally, if you are using wood products in your home – furniture, building supplies etc. – choose ones that are sustainably harvested. Look for certification from a group like the Forest Stewardship Council (FSC), an international organization that sets standards for responsible forest management. The FSC label tells you that particular product maintains FSC standards.

PRODUCTS AND SERVICES IN CALGARY

1. INTERIOR DESIGN

Housebrand helps homeowners create more livable, sustainable homes in their entirety. Designers evaluate the location of your home and its design, manage the construction or renovation, create the interior design and supply furniture.
www.housebrand.ca or 403-229-4330

Wharton Interior Design uses reduced energy use and healthy indoor air as the foundation for each design, creating convenient and durable indoor spaces.
www.whartoninteriors.com or 403-809-3923

Willow Studio provides interior design consulting services, furniture and accessories with the environment in mind. They emphasize accentuating what you already have, refurbishing and reupholstering.
www.willowstudio.ca or 403-230-9226

2. LIGHTING

Nemalux is a Calgary-based LED manufacturer that can create an energy-efficient, low-maintenance lighting plan that is right for you.
www.nemalux.com or 403-242-7475

3. SECOND-HAND FURNISHINGS

There are plenty of places to buy and donate second-hand furnishings, including Goodwill Industries of Alberta, Habitat for Humanity stores, Salvation Army Thrift stores and Value Village. The Consignment Gallery, Curious Cat Antiques and Honey B's offer vintage and quality second-hand home furnishings and accessories. And there are many local buy-and-sell websites.
www.buysell.com
calgary.kijiji.ca/
www.calgary.en.craigslist.ca
www.consignmentgallery.ca or 403-253-7880

www.curiouscatantiquesmall.com or 403-301-0157
www.goodwill.ab.ca (click Shopping)
www.habitatcalgary.ca (click Habitat ReStore)
www.honeybs.com or 403 283-0272
www.thriftstore.ca
www.usedcalgary.com
www.valuevillage.com (click Find a Store)

3. SUSTAINABLE BUILDING MATERIALS

Spruce up your home with ecocoat 100 per cent recycled latex paint, made in Calgary and available in 14 colours. Get it from Clean Calgary's The EcoStore or direct from the manufacturer.
www.cleancalgary.org or 403-230-1443
www.recyclepaint.com or 403-287-7726

Consider used building materials such as windows and doors, plumbing fixtures, kitchen cabinets and more from the Habitat for Humanity ReStore or Tim's Reusables.
www.calgaryhabitat.ca (click Habitat ReStore) or 403-291-6764
www.timsreusables.com or 403-276-1616

Riva's The Eco Store carries sustainable building products such as paint, flooring, insulation and countertops alongside environmentally responsible furnishings and mattresses.
www.rivasecostore.com or 403-452-1001

Ten Thousand Villages sells fairly traded home furnishings and accessories. Watch for their annual oriental rug sale that features handknotted carpets made by fairly paid adults.
www.tenthousandvillages.ca

RESOURCES

1. CHECKLISTS, FACT SHEETS AND GUIDES

Clean Calgary Association's *Green Building and Renovation Guide* provides information on environmentally sensitive cabinets, counters, carpets, paint, wallpaper and much more.
www.cleancalgary.org (under Guides) or 403-230-1443

Green Alberta is an online resource of green building products and materials that have been independently reviewed to assess their green attributes.
www.greenalberta.ca (under Database choose View the Database)

GREENGUARD Environmental Institute strives to improve public health and quality of life by improving indoor air quality. Use the GREENGUARD *Online Product Guide* to find indoor furnishings and finishing products that have been independently tested to ensure that their chemical and particle emissions meet acceptable indoor air quality standards.
www.greenguard.org

Reduce the toxic load in your home, room by room, by going through Toxic Nation's checklist.
www.toxicnation.ca (under Go Toxic Free select Toxic Nation Guides, then Toxic Nation – Make Your Home a Healthy Home!)

Find the top 25 indoor, pollution-reducing plants.
www.treecanada.ca (search "indoor plants" and choose the article "Indoor Plants for Health and Wellness")

> **Green Fact**
> A single spider plant can absorb 85 per cent of the formaldehyde in a room in just six hours!

2. THE SCOOP ON LIGHTING

The Government of Canada's Office of Energy Efficiency website explains everything you need to know about choosing energy-efficient lighting for both indoors and out.

www.oee.nrcan.gc.ca (under Personal Use click Residential, then Lighting)

REBATES AND GOVERNMENT PROGRAMS

1. NEW HOME REBATE

Buy a new EnerGuide-labelled single-family or multi-family row home and receive up to $10,000 in rebates from the government of Alberta!

www.climatechangecentral.com (under My Rebates)

8 TRANSPORTATION

GET ON THE GREEN TEAM

Another big contributor to our energy use – especially in Calgary – is how we get around. Reducing vehicle emissions is one way individuals can help reduce air pollution, which is a major health hazard. While the best way to reduce your reliance on a vehicle is to make sure you live close to work, play and the services you need, there are other things you can do to reduce your impact.

- Take transit to work and save money on car insurance, gas and parking.
- Check your tire pressure at least monthly. A car riding on under-inflated tires uses more fuel than one on properly inflated ones. Under-inflated tires also make your vehicle less safe.
- Change your oil regularly to promote better fuel efficiency and reduced emissions. Choosing the lowest multi-grade of oil recommended in your owner's manual can improve fuel efficiency, particularly when starting a cold engine!
- Don't idle. If you're stopped longer than 60 seconds (except in traffic) turn off your engine. Idling uses more fuel than turning the engine off and restarting it. Even in winter, today's computer-controlled, fuel-injected engines require no more than 60 seconds of warm up before driving.
- Use a block heater with an automatic timer. In temperatures below 0°C, block heaters can improve fuel economy by 10 per cent or more. Plug it in and set the timer for no more than two hours before you plan to depart.

PRODUCTS AND SERVICES IN CALGARY

1. ALTERNATIVE TRANSPORTATION

Calgary Transit keeps cars off the road by providing bus and CTrain service throughout the city. Route and schedule information is available at the click of a button.
www.calgarytransit.com or 403-262-1000

Calgary Transit has bike racks mounted on buses on certain bus routes, and bike locker rentals at certain CTrain stations. Learn more about bringing your bike on Calgary Transit at:
www.calgarytransit.com (search "bike") or 403-262-1000

Fun-time E-Scooters sells electric scooters and bicycles whose highly efficient electric motors use very little energy.
www.e-scooters.ca or 403-697-8835

The Good Life Community Bike Shop provides free space, tools and advice to help people build, repair and maintain their bicycles. They hold free repair workshops and also sell used bikes.
www.goodlifebikes.ca or 403-984-4727

Park 'n' Bike locations allow you to park for free, then walk, cycle or inline skate downtown. There are seven sites located five to eight kilometres from downtown Calgary.
www.calgary.ca (search "park and bike") or 3-1-1

Pete's Electric Bikes provides eco-sensitive transportation options in the form of electric bikes and scooters.
www.petesebikes.com or 403-444-9170

Power in Motion sells hybrid electric bicycles whose electric motors will give your pedal power a boost when you need it, and conversion kits that do that same for your existing conventional pedal bike.
www.powerinmotion.ca or 403-233-8841

2. CARBON OFF-SETTING

Air Canada and WestJet both offer options to carbon offset your flights. This is free on Westjet when you access their online booking through the Offsetters website. Air Canada offers the option to offset the effects of your flights for a fee through Zerofootprint.

www.aircanada.com (search "carbon offset")
www.offsetters.ca (under What We Do click WestJet)

3. CARPOOLING

Calgary Parking Authority's Carpool Parking program allocates 50 per cent of all new monthly parking contracts to those who will be driving with two or more occupants.

carpool.waitlist@calgaryparking.com

Carpool.ca is an online tool that helps match you with others who are interested in sharing the daily commute. If you have school-aged kids, see if their school participates in SchoolPool.ca, designed to connect carpooling families.

www.carpool.ca
www.schoolpool.ca

4. CAR SHARING

CATCO (Calgary Alternative Transportation Co-operative) promotes sustainable transportation in Calgary by operating a car sharing program. When you're a member, it's like you own a portion of a car, with your cost based on how much you use it.

www.calgarycarshare.ca

5. RETIRE YOUR RIDE

Retire Your Ride offers rewards such as cash and transit passes to get polluting vehicles off the road. Vehicle models from 1995 and older that are in running condition and have been registered and insured for the past six months are eligible. The parts are then recycled: fluids

are drained and reused or properly disposed; parts like batteries, tires and wheels are sold; and metals are crushed and recycled.
www.retireyourride.ca or 1-877-773-1996

> **Green Fact**
> Your vehicle produces 2.4 kilograms of carbon dioxide for every litre of fuel it uses.

RESOURCES

1. COMMUTING

Sustainable Alberta Association provides information, technical assistance and incentives for employers and employees to reduce the number of vehicles on the road.
www.calgarycommute.ca or 403-294-0904

2. FUEL-EFFICIENT VEHICLE INFO

Get advice on things to consider when shopping for a fuel-efficient vehicle, look at fuel-consumption ratings and comparisons, and see which vehicles are the most efficient in the annual ecoENERGY for Vehicles Awards.
www.oee.nrcan.gc.ca (under Personal Use click Transportation)

If you're in the market for a hybrid vehicle, you'll want to find out more about hybrid technology, how a hybrid might fit into your lifestyle, see reviews and view comments from current hybrid owners.
www.hybridcenter.org

3. IDLING AWARENESS

Start a public awareness campaign to reduce vehicle idling. Start by downloading posters, brochures and street signs.
www.oee.nrcan.gc.ca (under Personal Use click Transportation, then Idling and look under Building Your Own Campaign and Resources)

Encourage others to stop idling by getting a bumper sticker from Pink Gazelle Productions.
www.pinkgazelle.com (click Buy a Vehicle Decal)

> **Green Fact**
> If every Canadian driver avoided 3 minutes of idling each day, in a year we would save 630 million litres of fuel, over 1.4 million tonnes of carbon dioxide emissions, and $630-million in fuel costs (assuming fuel costs of $1/L). See www.oee.nrcan.gc.ca (under Personal Use, click Transportation, then Idling)

4. RECYCLING VEHICLE WASTES

Alberta Used Oil Management Association collects and recycles used oil, filters and containers from do-it-yourselfers, farmers and small businesses. Find the nearest drop-off location at:
www.usedoilrecycling.com or 1-888-922-2298

5. GET ON THE WALKING SCHOOL BUS

Start a walking or cycling school bus, where neighbourhood children walk or cycle to and from school together under the supervision of an adult. This reduces traffic congestion and vehicle emissions around schools and promotes physical activity, community interaction and safety.
www.saferoutestoschool.ca or 1-877-533-4098

REBATES AND GOVERNMENT PROGRAMS

1. HYBRID INSURANCE REBATE

Desjardins General Insurance offers 10 per cent off insurance premiums for hybrid vehicles.
www.desjardinsgeneralinsurance.com (search "hybrid")

> **Eco-words**
>
> *Hybrid vehicle:* a vehicle that combines a regular internal combustion engine with an electric motor. The electric motor can power a variety of functions depending on the design of the vehicle, including braking, idling and driving. The electric motor provides higher fuel efficiency and reduced emissions, while the conventional engine supplies predictable power and driving range.

2. TRANSIT PASS DEDUCTION

Save around $150 at tax time: the federal government offers a non-refundable tax credit to cover the cost of a monthly public transit pass. Keep receipts and the passes to claim the deduction on your tax return.

www.cra-arc.gc.ca (search "public transit pass")

> **Green Fact**
>
> The CTrain was the first public Light Rail Transit system in North America to run solely on wind-generated electricity.

FOOD AND DRINK

GET ON THE GREEN TEAM

Choosing food that is more environmentally responsible is one of the most pleasurable ways to live green. Know the pleasure of growing your own organic vegetables or getting to know your local farmers and producers. Community gardens and farmers markets are great places to meet people and share community spirit.

Eating more local food has become important to many people, as demonstrated by the popularity of the hundred-mile diet. The age-old arts of canning, freezing and dehydrating produce are making a comeback. But how can you make the most environmentally sensitive food choices when local isn't an option? Take the Locavore Pledge:

- If not FROM BACKYARD, then choose Locally Produced.
- If not LOCALLY PRODUCED, then choose Organic.
- If not ORGANIC, then choose Family Farm.
- If not FAMILY FARM, then choose Local Business.
- If not a LOCAL BUSINESS, then choose Fair Trade.

While you are thinking "local" consumption, consider the water you drink, too. Drink water from the tap rather than bottled water, which could be coming from your local municipal water source anyway. Drinking tap water reduces pollution produced when plastic bottles are manufactured, fossil fuels used when water is shipped, waste in the form of water bottles, and pollution when the bottles are buried or burned.

If you need water to go, fill up a stainless steel or glass container rather than polycarbonate (#7 plastic) water bottles. These plastic water bottles leach bisphenol A (BPA), a hormone disrupter. On June

26, 2009, the Government of Canada announced that it will move forward with regulations to prohibit the advertisement, sale and importation of polycarbonate plastic baby bottles that contain bisphenol A. You may wish to distance yourself and your family from this substance.

> **Eco-words**
> *Calgary Dollars:* The Arusha Centre's Calgary Dollars is a grassroots local currency. It encourages the consumption of local products and services to strengthen our local economy and build community.

PRODUCTS AND SERVICES IN CALGARY

1. FARMERS MARKETS

Blackfoot Farmers Market is open on Saturdays, May through October.
www.blackfootfarmersmarket.com or 5600 11 Street SE

Calgary Farmers Market is open year round on Fridays, Saturdays and Sundays.
www.calgaryfarmersmarket.ca or Currie Barracks Hangar #6, 4421 Quesnay Wood Drive SW

Crossroads Market is Calgary's largest year-round indoor/outdoor market. Open Fridays, Saturdays and Sundays.
www.crossroadsmarket.ca or 1235 26 Avenue SE

Grassroots Northland Farmers Market is open Tuesday afternoons from June to September.
www.skatersfoundation.com or at Northland Village Mall, 5111 Northland Drive NW

Hillhurst-Sunnyside Farmers Market is open seasonally Wednesday afternoons from June to October at the Hillhurst-Sunnyside Community Association, 1320 5 Avenue NW.
www.hillhurstsunnysidecommunity.com (under Markets) or 403-283-0554

McKenzie Towne Farmers Market is open Thursday afternoons June through September in the parking lot of McKenzie Towne Hall, 40 McKenzie Towne Boulevard SE.
pro.mctowne@shaw.ca or 403-781-6612 ext. 2

Millarville Market is open Saturday mornings from 8:30 until noon, June through October, at the Millarville Racing & Agriculture Society just southwest of Calgary.
www.millarville-ab.com/market.html or 403-931-2404

South Fish Creek Farmers Market is open Friday afternoons from mid-June until mid-September in the south parking lot of the South Fish Creek Complex, 333 Shawville Boulevard SE.
www.sfcra.com or 403-201-8652

Sweetgrass Market is Calgary's first year-round, week-long urban market. It emphasizes the value of local food and its producers.
www.sweetgrassmarket.ca or 873 85 Street SW

2. LOCAL ANIMAL PRODUCTS

Country Natural Foods delivers farm-raised chicken, turkey, lamb, beef, pork, bison and elk right to your door. The animals are naturally raised with no antibiotics, hormones or animal by-products in their feed.
www.countrynaturalfoods.com or 1-888-376-4325

Golden Lane honey produces honey from their beehives northeast of Calgary, but growing demand for their product means they now buy from other Canadian beekeepers. Most of their honey is produced within 20 kilometres of the city. Find them at the Calgary Farmers Market.
www.goldenlanefoods.com or 403-287-7277

Guitton Farms raises certified organic beef that can be picked up from their Claresholm area farm or delivered to Calgary. Check the website for other retail locations in the city.
www.guittonfarms.com or 403-625-4732

Hoven Farms certified organic Alberta beef is available year round at the Calgary Farmers Market and several other Calgary locations.
www.hovenfarms.com or 1-800-311-2333

Valta Bison Farms sells ethically produced food that comes directly from Alberta farms: a wide variety of meats, cheeses and preserves. They have three locations in Calgary: the Calgary Farmers Market, Sweetgrass Market and in Ramsay at 703 23 Avenue SE.
www.valtabison.com or 403-650-2425

3. LOCAL BAKERIES

A Ladybug Bakery & Café now has two locations where pastry artists create decadent treats using organic flour, eggs and milk. The new café location on Aspen Stone Boulevard SW serves soups, sandwiches and panini made from local meat and vegetables. Original location is at the Calgary Farmers Market.
403-287-1137 or at the Calgary Farmers Market

Earth's Oven specializes in gluten-free, easily digestible baked goods. Many ingredients are organic and all are preservative free.
www.earthsoven.com or 403-686-4810

Lakeview Bakery produces specialty baked goods including organic, wheat-free, yeast-free and sugar-free.
www.organicbaking.com or 403-246-6127

Prairie Mill Bread Co. bakes pioneer breads using 100 per cent organic prairie wheat. No preservatives or oils are added.
www.prairiemillbread.com or 403-282-6455

4. LOCAL DELIVERY

Small Potatoes Urban Delivery (SPUD) and Farm Fresh Organics deliver organic produce and groceries to your door.
www.spud.ca or 403-615-3663
www.freshorganics.ca or 403-210-3700

Spoon Fed delivers fresh soup to your door weekly. All the soups are made from fresh, seasonal, local ingredients, and they accept Calgary's local currency (Calgary Dollars) as payment.
www.spoonfedsoup.com or 403-801-1394

5. LOCAL "GREEN" GROCERS

Amaranth Whole Foods promotes healthy foods, nutritional awareness and environmental responsibility.
www.amaranthfoods.ca or 7 Arbour Lake Drive NW

Blush Lane Organic Market sells only certified organic produce and supports local growers first.
www.blushlane.com or 3000, 10 Aspen Stone Boulevard SW and the Calgary Farmers' Market

Community Natural Foods is a locally owned source for organic food and sustainable lifestyle products. They were Calgary's first grocer to stop providing single-use plastic bags.
www.communitynaturalfoods.com or 1304 10 Avenue SW and 202 61 Avenue SW

Planet Organic carries fresh organic produce and gourmet groceries.
www.planetorganic.ca or 10233 Elbow Drive SW and 110 4625 Varsity Drive NW

Sunnyside Market is a family-owned whole foods market specializing in local, organic and natural products.
www.sunnysidemarket.ca or #10, 338 10 Street NW

The Light Cellar sells raw, organic, fair-traded bulk foods such as dried fruits, nuts, seeds and spices. They also hold workshops and classes on plant-based cuisine.
www.thelightcellar.ca or 6326 Bowness Road NW

6. LOCAL, ORGANIC TAKE-OUT

Red Tree creates gourmet food to go from mostly local or organic ingredients.
www.redtreecatering.com or 403-242-3246

Forage makes take-away food from local ingredients. Find fresh daily specials, ready-to-heat and frozen meals and local specialty groceries.
www.foragefoods.com or 403-269-6551

Route 40 Soup Co. makes their amazing soups with local ingredients from producers that share their passion for taste and quality. They also know that by using locally produced food they're putting dollars back into their community. Find soups to take home at their Turner Valley location or at the Millarville Farmers Market.
www.route40.ca or 403-933-7676

7. LOCAL VEGETABLE GROWERS

Innisfail Growers is comprised of five farm families who cooperatively bring their produce to local markets, including several in the Calgary area. Together they produce asparagus, tomatoes, flowers, potatoes, carrots and much more.
www.innisfailgrowers.com or at Calgary Farmers Market, Grassroots Northland Farmers Market and others

Lund's Organic Farm sells vegetables that are the best of both worlds: local and organic. Certified organic since 1993, they grow a variety of vegetables with the emphasis on cold-weather crops such as carrots, spinach, lettuce and potatoes.
www.lundsorganic.com or at the Calgary Farmers Market

Thompson Small Farm is a CSA (Community Supported Agriculture) farm just east of Beiseker. Members share in the harvest from crops grown, mostly cold-weather vegetables like carrots, onions, spinach, peas, beans, potatoes and more.
wrightjdigby@live.com or 403-572-3032

8. LOCAL WINERIES ... AND A MEADERY

Chinook Arch Meadery turns Chinook Arch honey into the oldest alcoholic beverage in the world: mead. With names like Ginger Snapped and Bodacious Black Currant, who wouldn't want to try

some? Find both the mead and the honey at their location southwest of Okotoks or at the Millarville Market.
www.chinookhoney.com or 403-995-0830

enSanté Winery is Alberta's only organic cottage winery. These wines are made from organic raspberries, cherries, rhubarb, saskatoons, apples and honey.
www.ensantewinery.com or 780-657-2275

Field Stone Fruit Wines are vinted from raspberries, strawberries, wild black cherries and saskatoons grown just east of Calgary. Visit their on-site wine store May through September, or visit them at the Calgary Farmers Market.
www.fieldstonefruitwines.com or 403-934-2749

9. SOWING SEEDS

Seedy Saturday is about saving, swapping, buying and selling seeds that are open-pollinated and have not been genetically modified. Watch for it every spring.
www.calhort.org (under Gardening Resources click on Community Gardens)

RESOURCES

1. COMMUNITY GARDENS

The Community Garden Resource Network provides information on community gardens in Calgary, as well as resources for starting and maintaining a community garden. Grow your own local organic produce!
www.calhort.org (under Gardening Resources click on Community Gardens)

The Garden Path Society has Calgary's largest community garden, with 103 individual plots and almost half an acre of community beds.
www.calgarygardenpath.ca

2. FARMERS MARKETS

Farmers markets can come and go. For current listings of markets check out the Alberta Farmers Market Association.
www.albertamarkets.com

3. ORGANIC PRODUCTION GUIDES AND REGULATIONS

> **Eco-words**
> *Organic:* crops that are grown without conventional pesticides, artificial fertilizers, human waste or sewage sludge, that are processed without ionizing radiation or food additives, and that are not genetically modified. Organic meat and animal products are reared without the routine use of antibiotics and growth hormones.

Organic product regulations came into effect in Canada on June 30, 2009. This means that food products that meet National Organic Standards can display the Biologique Canada Organic designation and logo.
www4.agr.gc.ca (in sidebar, click "Subjects A–Z," then "Organic Food")

Canada's Seafood Guide helps Canadians purchase sustainable seafood from Canada and abroad. The guide lists the most ocean-friendly seafood in a handy, downloadable wallet guide. It is compiled by Sustainable Seafood Canada, a coalition of conservation organizations including the David Suzuki Foundation and the Sierra Club of Canada.
www.seachoice.org

Learn about the pesticides on your produce from the Environmental Working Group. There's even a handy wallet guide or iPhone app listing which produce has the highest and lowest pesticide loads.
www.foodnews.org/walletguide.php

4. SLOW FOOD CALGARY

Slow Food seeks to connect local, sustainable food producers and processors with consumers. The idea is to protect regional cuisines, ingredients and small suppliers from industrialization.
www.slowfoodcalgary.ca

5. U-PICK

You'll know it's fresh if you pick it yourself. Check out a U-pick farm near you.
www.albertafarmfresh.com

6. VEGETARIAN RESTAURANTS

The Calgary Vegetarian Society provides a list of vegetarian restaurants and local vegetarian events.
www.calgaryveg.com

> **Green Fact**
> Meat production requires more water than raising crops. Producing 275 grams (10 oz.) of beef uses 85 times more water than growing the same amount of potatoes.

WASTE AND RECYCLING

GET ON THE GREEN TEAM

Everyone has heard about the three Rs: Reduce, Reuse, Recycle. But have you ever thought about what each R really means, and how you can practise all three in your everyday life?

"Reduce" simply means use and buy less. Ask yourself the following questions when you are considering buying something new:

- Do you need to own it? Can you borrow it from a friend? Share it with someone else? Rent things that you don't use very often.
- If you must buy, does it have to be new? Can you find it second hand?
- How is it packaged? Buy things with less packaging; avoid individually wrapped food and other items. Buy in bulk when it makes sense.

> **Green Fact**
> Packaging makes up about half of our garbage by volume, one-third by weight.

"Reuse" means don't buy that new thing, but instead reuse what you already have or what someone else has. Here are some ways to reuse:

- Buy items second hand.
- Avoid disposable items: everything from batteries to coffee cups to razors.
- Donate, give away or share items you don't need.
- Repair what you have rather than buying a new one.

- Find a way to use the item other than what it was originally intended for.
- Buy products that will last.

"Recycle" means find out where products and packaging came from before you buy and know that they can be recycled when you are done with them.

- Look for recycled content in everything you buy, from paper to aluminum foil to toothbrushes.
- Ensure the goods you purchase are recyclable.
- Support local recycling programs.
- Compost.

There is a fourth R. "Rethink" means ask questions about the goods you buy and make sure you read labels. Here are some questions you can ask yourself:

- Who produced the product? Would they have been fairly compensated?
- What is it made of? Is it renewable, recycled, recyclable? Is it made from something toxic?
- Where was it made? How far has it had to travel to get to me, the consumer?
- How was it made? How many resources and additives were used to make it, and how much waste was created? Was it produced in a sustainable manner?
- Why do I need it?

Beyond the Rs, following are some more tips about waste and recyling:

- About 35 per cent of household waste could be composted and kept out of the landfill. (See Chapter 5 for composting tips and resources.)

- Recycle right: clean recyclable cans, bottles and jars in old dishwashing water. If you must use fresh tap water, use cold, not hot.
- Put a No Flyers sticker on your mailbox.
- Use less paper by printing double-sided and buying post-consumer recycled content paper.
- Take past-date medications to the pharmacy for disposal rather than flushing them down the toilet or putting them in the garbage. Medications contaminate groundwater and should be disposed of professionally.

Green Fact
The City of Calgary collects methane produced by decomposing organics in the East Calgary and Shepard landfills and uses it to generate electricity.

PRODUCTS AND SERVICES IN CALGARY

1. BOTTLE DEPOTS

Find locations of the nearest bottle depot and information on what they accept and how much they pay from the Alberta Beverage Container Recycling Corporation:
www.abcrc.com or 403-264-0170

2. CHEMICAL DISPOSAL

Take household chemicals – like paint, cleaning products, antifreeze, pesticides and nail polish – to one of The City of Calgary's eight, year-round chemical drop-off depots for free, safe disposal. See Appendix for more information.
www.calgary.ca/waste (under Quick Links choose Household Chemical Drop-off Program) or 3-1-1

3. CONNECT AND RECYCLE

BorrowMe offers alternatives to traditional consumption by linking people to share, rent, gift or sell their possessions. It also has a component to help establish lending circles for groups, clubs, neighbours and friends.
www.borrowme.com

Join Freecycle. It's a free, non-profit group. Through Freecycle, people give away items they no longer need and find things they do need that others are giving away. This helps to keep things out of the landfill and helps keep us from having to purchase new items.
www.groups.yahoo.com/group/freecyclecalgary

> **Green Fact**
> Airdrie and Cochrane residents are limited to two bags of trash each week!

> **Green Fact**
> The following items are accepted in The City of Calgary's blue carts and depots:
> plastics labelled with recycling codes 1 through 7, newspapers, magazines, mixed paper, cardboard, milk containers, metal cans, glass, plastic bags

4. ELECTRONICS, PAINT AND TIRE RECYCLING

Alberta Recycling Management Authority can help you to find electronics, paint and tire recycling collection sites near you.
www.albertarecycling.ca or 1-888-999-8762

Cell phones can be recycled at the Calgary Zoo. Take them to the admission gates and the zoo will receive a refund from ECO-CELL.
www.calgaryzoo.org or 403-232-9300

The City of Calgary's Electronic Recycling Depot Locations Map will also help you find recycling locations. See Appendix for more information.
www.calgary.ca/waste (under Quick Links click on Electronics Recycling Locations) or 3-1-1

Clean Calgary Association accepts cell phones and PDAs for recycling. They get $1 for every unit they turn in to Pitch-In Canada.
www.cleancalgary.org or 403-230-1443

Donate your old computer to Computers for Schools (CFS), a program sponsored by Industry Canada to refurbish computers and related equipment and distribute them to Canadian schools and libraries.
www.ic.gc.ca/eic/site/cfs-ope.nsf/Intro

Rechargeable batteries for cellular and cordless phones, power tools, laptops, video and digital cameras, toys and more can be recycled through the Rechargeable Battery Recycling Corporation. They reclaim materials such as nickel, iron, cadmium, lead and cobalt, keeping them out of the landfill. They also recycle cell phones. Find a drop-off location near you.
www.rbrc.org (click Recycle Now)

Recycle Logic is approved by the Alberta Recycling Management Authority to process electronic waste such as televisions, computers and printers. They also accept many other electronic devices.
www.recycle-logic.com or 403-266-1012

Technotrash accepts all electronics – even Christmas lights – for recycling and is authorized by the Alberta Recycling Management Authority.
www.technotrash.ca or 403-265-2332

5. INK AND TONER RECYCLING

Printer ink and toner cartridges can be reused and recycled. Find a dealer near you in Clean Calgary's *Reuse and Recycling Guide* or The City of Calgary's Recycling Opportunities List.
www.calgary.ca/waste (click Downloadable Files)
www.cleancalgary.org (click Guides)

6. LIGHT BULB AND BATTERY RECYCLING

Compact fluorescent light bulbs (CFLs) and single-use batteries are accepted for recycling at IKEA. Drop them off in the bins next to the Returns & Exchanges desk.
www.ikea.ca or 403-273-4338

Mountain Equipment Co-op accepts single-use batteries, 100 per cent polyester clothing and rechargeable batteries for recycling.
www.mec.ca or 403-269-2420

RONA and Home Depot also accept CFLs for recycling.
www.rona.ca
www.homedepot.ca

7. RECYCLING SERVICES

As of mid-2009, all single-family dwellings in Calgary have curbside recycling. Each residence has its own blue cart to be filled with recyclables and put out with the weekly garbage. Community recycling depots remain in place for multi-family dwellings, with improvements so they require no sorting and accept plastic. Visit the following website for details.
www.calgary.ca/waste or 3-1-1
www.calgary.ca/waste (under Quick Links click Recycling Depots) or 3-1-1 or see Appendix.

Clean Calgary's *Reuse and Recycling Guide* keeps an up-to-date list of independent recycling companies.
www.cleancalgary.org (under Guides) or 403-230-1443

If you want recycling pick up at your condominium, townhouse or apartment, contact one of these independent recycling companies:

Condo Recycling Solutions
www.condorecycling.com or 403-680-7557

Greenway Recycling
recycle2@telusplanet.net or 403-263-9025

Recycle Blue
403-230-1030

Residential Recycling
403-245-4451

The Curbside Recycling Association of Southern Alberta can also help you find recycling companies.
crasa@shawbiz.ca or 403-276-7555

> **Green Fact**
> The City of Calgary estimates that in 2008, 20 per cent of the city's waste was kept out of landfills by recycling and other waste diversion programs. With the implementation of the curbside recycling program in 2009, the city estimates that proportion will double to 40 per cent. The city's goal is to divert 80 per cent of all waste from landfills by 2020.

8. STOP JUNK MAIL

Stop junk mail with a No Flyers sticker from Clean Calgary Association. It only costs $2 and is available from Clean Calgary's The EcoStore and other locations throughout the city.
www.cleancalgary.org (search "no flyers")

At the Red Dot Campaign you can purchase or download a No Junk Mail sign, find a prewritten letter to Canada Post if your sign gets ignored, and more.
www.reddotcampaign.ca

RESOURCES

1. CHRISTMAS TREES AND PUMPKINS

Recycle your Christmas tree, bagged leaves and pumpkins free of charge at one of The City of Calgary's many seasonal drop-off locations or at any of the three landfills. The East Calgary Landfill accepts leaves and pumpkins without charge year round.
www.calgary.ca (search "leaf and pumpkin")

2. COMPOSTING (SEE ALSO CHAPTER 5)

The City of Calgary has a great guide to get you started composting.
www.calgary.ca (search "what is composting" and select the *Composting in Calgary* pdf)

Learn how to compost by taking a workshop from Clean Calgary Association.
www.cleancalgary.org or 403-230-1443

Clean Calgary Association's Compost Hotline is the perfect medium for citizens to ask those burning composting questions and get answers. Call them with your composting questions, troubles and success stories.
403-998-8846

Compost pet waste and meat products with The Digester, available from Clean Calgary's The EcoStore.
www.cleancalgary.org or 403-230-1443

Learn how to employ worms to eat your organic waste. The Province of Alberta has a guide called *Taking Action through Vermicomposting to Reduce Kitchen Waste*.
www.environment.gov.ab.ca/info/library/6188.pdf

Vermicomposting allows you to let worms eat your organic waste. It's perfect for people who can't compost outdoors. Bins and worms are available from Clean Calgary's The EcoStore and Worms@Work.
www.cleancalgary.org or 403-230-1443
www.wormsatwork.com or 403-730-3405

Eco-words

Vermicomposting: feeding your organic garbage – like vegetable scraps, coffee grounds and eggshells – to worms, which break it down into fertilizer you can use in your house plant soils and garden.

Green Fact

Garbage only decomposes with the help of water and oxygen, but since the garbage in landfills is tightly compressed, water and oxygen can't get in to help break it down. Eventually landfill garbage does break down, but the by-products are methane (a greenhouse gas 20 times stronger than carbon dioxide) and leachate (a toxic garbage soup). Our landfills produce 38 per cent of Canada's methane emissions.

$ave Green

The Earth Machine Composter is only $23.58, thanks to a subsidy from The City of Calgary. Get one at Clean Calgary's The EcoStore, or have it shipped to your door from The City of Calgary Online Store.
www.cleancalgary.org or 403-230-1443
www.calgaryonlinestore.com (search "composter")

3. GET OFF THE LIST

If you are an Air Miles collector, have your name removed from their mail distribution lists by contacting them.
1-888-247-6453 or privacyoffice@airmiles.ca

Register online for the Canadian Marketing Association's Do Not Mail/Do Not Call program.
www.the-cma.org/consumer/donotcontact

Check with your service providers to see if they can distribute your bills online. Or register with epost, Canada Post's free online bill delivery service. epost allows you to receive, store and pay over 200

types of bills electronically. They even have a Green Calculator that tells you how many trees you will save.
www.epost.ca

4. GUIDES AND LISTS

The City of Calgary's A–Z Recycling Disposal Directory can also direct you to where to recycle goods.
www.calgary.ca/waste (under Quick Links)

Clean Calgary's *Reuse and Recycling Guide* is a comprehensive resource for where you can reuse and recycle everything in Calgary, from appliances to printer cartridges to yard waste. Use it to find second-hand goods or to donate things you are no longer using.
www.cleancalgary.org (click on Guides)

"The Green Guide" appears every Friday in the *Calgary Herald*'s "Real Life" section and has tons of handy information, including Recycling 101 and the *Calgary Recycling Guide*.
www.calgaryherald.com/life/green-guide

CHILDREN

GET ON THE GREEN TEAM

Because children consume more food and drink and breathe more per unit of body weight than adults, they are more vulnerable to toxic chemicals in the environment. Furthermore, they could be exposed to toxic chemicals at critical times in their development, resulting in more severe repercussions.

Be aware of the chemicals in the lotions, oils and powders you put on your child's skin. Many conventional products are made from petroleum and other ingredients that are known to be irritating and harmful.

Disposable diapers also contain potentially irritating perfumes and are heavily bleached. And of course, disposable diapers end up in the landfill. Even biodegradable diapers don't break down quickly in landfills, because of the lack of oxygen and water. Consider alternatives like reusable diapers and diaper services.

Baby bottles and sippy cups made from #7 plastic can leach bisphenol A, a known hormone disrupter that has been linked to cancer, impaired immune function, early onset of puberty, obesity, diabetes and hyperactivity. Consider glass or #1 or #2 non-leaching plastic. (Note that the Government of Canada is moving forward with regulations to prohibit the advertisement, sale and importation of polycarbonate plastic baby bottles that contain bisphenol A.)

PRODUCTS AND SERVICES IN CALGARY

1. CONSIGNMENT STORES FOR KIDS AND EXPECTING MOMS

Lullaby Lane aims to be the one place moms can find everything they need. Find quality second-hand children's and maternity clothing, toys and accessories.
www.lullabylane.org or 403-264-2625

Kids on the Grow Konsignment has clothes, bedding, toys, cribs and so much more for your newborn to 14-year-old. Every item is inspected to ensure it meets the most recent safety regulations.
403-590-1536

Sproutz Kidz resells brand-name children's clothing, furniture and accessories, emphasizing that consignment shopping is earth friendly and great value.
www.sproutzkidz.com or 403-242-9877

2. DIAPER SERVICES

Happy Nappy Diaper Service exchanges your dirty cloth diapers for fresh clean ones every week.
www.happynappy.ca or 403-281-6100

3. GREEN PARENTING COURSES

Green Plan-It offers eco-education to help parents and children make more environmentally sensitive choices. Choose from classes like Eco-Baby 101 and Goin' Green for Parents.
www.greenplan-it.ca

4. SHOPPING FOR THE SUSTAINABLE BABY

Babes in Arms specializes in babywearing and cloth diapering products, and it also carries organic and sustainable options for toys, nursing wear and skin care. In addition, it has related books for parents and babies.
www.babesinarms.ca or 403-835-4614

Barakat Ecostore offers goods that are fairly traded and kind to the earth such as hemp clothing, bamboo towels, cleaning products, body care products, baby products, organic snacks and teas and more.
www.barakatecostore.com

Claudia's Choices sells several lines of cloth diapers, natural baby skin care, environmentally friendly laundry options and more.
www.claudiaschoices.ca or 1-888-613-9274

The Cloth Caboose Baby Store offers a variety of cloth diapers and other natural products like skin care products and organic cotton toys.
www.clothcaboose.ca or 403-730-6564

ComfortCotton.com is a Calgary-based online store dedicated to offering organic cotton bedding, diapers and clothes for your children.
www.comfortcotton.com or 403-285-4853

Fibres of Life sells bright, durable and fairly traded felt handicrafts from Nepal. For children they feature adorable slippers, booties, hats and toys. Check out the finger puppets!
www.fibresoflife.com

Growing up Organic is an online store specializing in certified organic, fair-trade cotton baby gear, from diapers to bedding, nursing pillows to teething toys.
www.growinguporganic.ca or 403-618-1780

Nature Babies is all about green living for little ones. They have a broad range of products – from apparel to bedding to eating – all in one place, so it's easy for parents who have decided to go green to get what they need.
403-988-6659

organicKidz is a Calgary company that sells stainless steel baby bottles and sippy cups. The products are light, dishwasher safe and of course do not contain BPA, phthalates, PVC or lead.
www.organickidz.ca or 403-201-2585

Riva's The Eco Store carries many healthy, sustainable baby products such as diapers, bibs, bedding, bottles, clothes and much more.
www.rivasecostore.com or 403-452-1001

Diaper Depot has been manufacturing Gabby's cloth diapers in Calgary for 20 years. See their line made from organic cotton.
www.gabbys.net or 1-888-442-2297

5. EMPOWERING THROUGH THEATRE

Evergreen Theatre produces musical theatre for youth to inspire and empower them to keep the world green. Performances feature the company's original music and are linked to school curriculum.
www.evergreentheatre.com or 403-228-1384

RESOURCES

1. ALTERNATIVE WAYS TO GET TO SCHOOL

Start a walking or cycling school bus, where neighbourhood children walk or cycle to and from school together under the supervision of an adult. This activity reduces traffic congestion and vehicle emissions around schools and promotes physical activity, community interaction and safety.
www.saferoutestoschool.ca

If you have school-aged kids, see if their schools participate in SchoolPool.ca, designed to connect commuting families. School-pooling reduces traffic and vehicle emissions around schools.
www.schoolpool.ca

2. GUIDES, CHECKLISTS AND WATCHDOGS

Environmental Working Group has a whole section on their website dedicated to parents that is full of tips, tools and guides regarding product safety for the whole family. See especially

their *Guide to Children's Personal Care Products* and *Guide to Feeding Baby*.
www.ewg.org/forparents

The non-profit Environmental Working Group's free Skin Deep database evaluates the hazard level of baby care products, from diaper cream to toothpaste to wipes.
www.cosmeticsdatabase.com (under Baby Care)

The Guide to Less Toxic Products is a Canadian source for information about the risks and alternatives to common baby care products.
www.lesstoxicguide.ca (choose Baby Care)

> **Green Fact**
>
> Phthalates are plastic-softeners that are often used in children's toys. They have been banned in the EU because they are suspected to cause hormone irregularities and cancer. These chemicals are not banned in children's toys in Canada; the bill introduced to ban them (Bill C-307) languished when the October 2008 election was called. To find out more about Bill C-307, see www.toxicnation.ca/mediacentre/readnews/260.

REBATES AND GOVERNMENT PROGRAMS

1. MAYOR'S ENVIRONMENTAL EXPO

The Mayor's Environment Expo is held annually in June in celebration of National Environment Week. This event offers interactive workshops, live theatre, student presentations and over 50 exhibits to showcase how every small step we take helps decrease Calgary's ecological footprint. Ensure your child's class is attending.
www.calgary.ca (search "Mayor's Environment Expo")

$ave Green
The Jack Leslie Youth Environment Grant funds youth-initiated projects that have a positive impact on the environment in Calgary. One $500 and two $1,000 grants are available annually.
www.cyfc.ca (under Involve choose Programs and look under Funding Opportunities) or 403-266-5448

12
CLOTHING AND PERSONAL CARE

GET ON THE GREEN TEAM

Often we forget that the skin is our largest organ, and that one of the primary ways we are exposed to toxins is through our skin. Therefore, it's important to look for cosmetics and personal care products with fewer harmful chemicals. Watch the label for ingredients like sodium lauryl/laureth sulfate, triclosan, propylene glycol, petrolatum, parabens and anything labelled as "fragrance" or "colour," all of which have questionable effects on human health. Propylene glycol, for example, is a very common ingredient that is recognized as a neurotoxin by the National Institute for Occupational Health & Safety in the U.S. and is known to cause contact dermatitis, kidney damage and liver abnormalities.

As well, the chemicals in our personal care products that get washed down the drain are not removed from waste water. Chemicals like the anti-bacterial triclosan – found in many products, including hand soap and toothpaste – make their way through the waste-water treatment system and back into water sources. Triclosan breaks down into chemicals that are toxic to aquatic life: it causes the same type of thyroid disruption in frogs as it does in humans.

Our clothing too is in constant contact with the skin. Pesticide residues, dyes and chemicals from dryer sheets can all be absorbed through the skin, though the exposure decreases every time you wash a garment. When clothes shopping, look for certified organic cotton products or more eco-friendly fibres. Hemp, bamboo, soy and woollen fibres can all be sustainably harvested.

Alternatively, choose not to buy new clothes! A fun way to reuse clothing is to have a clothing swap: invite your friends over, telling them to bring some gently used clothes. Pile all the clothes in the middle of the floor and take turns picking new-to-you items.

Regardless of whether a clothing item is used or new, it's always best to choose clothes that don't have to be dry cleaned. Conventional dry cleaners most often use perchloroethylene (perc), a chlorinated solvent that is a possible human carcinogen. Minimize dry cleaning and ask your dry cleaner about wet cleaning, carbon dioxide cleaning or other non-perc alternatives.

> **Eco-words**
> *Organic cotton* is grown without pesticides from plants that are not genetically modified. Conventional cotton production uses more chemicals than any other crop type and accounts for 25 per cent of the world's pesticide use.

PRODUCTS AND SERVICES IN CALGARY

1. EARTH-FRIENDLY NEW-CLOTHING OPTIONS

American Apparel is all made in Los Angeles by fairly paid workers. Moreover, they have a line of 100 per cent organic cotton basics: thongs, tanks and tees.
www.americanapparel.net or 403-410-7450

Barakat Ecostore offers goods that are fairly traded and kind to the earth, such as hemp clothing, bamboo towels, cleaning products, body care products, baby products, organic snacks and teas, and more.
www.barakatecostore.com

The Earth Collection uses natural fibres such as organic cotton, silk, linen, ramie (a linen-like fiber) and hemp to create classic, durable clothing for men, women and children. EU Eco-Label, which

certification means the product has been verified by a third party to comply with strict ecological and performance criteria (equivalent to Canada's EcoLogo).
www.theearthcollection.com or 403-270-8858

Elements – The Patagonia Store carries clothing that is increasingly made from recycled and recyclable polyester, organic cotton, hemp, organic wool and chlorine-free wool. Plus, their Common Threads Garment Recycling Program allows customers to return worn-out clothes from several Patagonia lines to be recycled into new clothes!
www.patagonia.com or 403-266-6463

Eleven Eleven Boutique in Kensington specializes in colourful dresses made from Indian sari fabric and reclaimed jewellery by local designers, as well as carefully selected vintage clothing.
403-452-5285

Fibres of Life sells whimsical, contemporary, fairly traded felt handicrafts created by Nepalese artisans. See their funky bags, slippers, hats and more at the Calgary Farmers Market.
www.fibresoflife.com

Gravity Pope carries several lines of eco-friendly footwear. More shoemakers are incorporating recycled materials into their footwear and are using more earth-friendly materials like hemp canvas, felt and water-based glues.
www.gravitypope.com or 403-209-0961

Matt & Nat vegan handbags are designed in Canada. The linings in their bags are made entirely from recycled plastic bottles.
www.mattandnat.com or 403-410-9236

Mountain Equipment Co-op (MEC) has many earth-friendly clothing options made from organic cotton and recycled polyester fibres. Canadian and member owned, MEC is committed to ethical sourcing and sustainability.
www.mec.ca or 403-269-2420

Riva's The Eco Store has everything green you could want to put on your body. They carry stylish clothes and shoes made from new organic and recycled fibres, as well as several lines of organic skincare and beauty products.
www.rivasecostore.com or 403-452-1001

2. SECOND-HAND CLOTHING

Cat's Eye Vintage is a huge space filled with clothing and accessories sorted by decade. Be prepared to spend some time perusing their vast collection. Voted first runner-up for Calgary's Best Consignment/Vintage Clothing Store by FFWD *Weekly* readers in 2008 and 2009.
403-640-4090

Changes Consignment Clothing Co. has stylists to help you create a consignment wardrobe that is right for you.
www.changesclothing.com or 403-240-3392

Danielle's Consignment Boutique sells hand-picked, high-end clothing, shoes and accessories
www.daniellesconsignment.com or 403-244-4752

Krysta Taylor Style Consulting provides wardrobe renovations, giving you a new look with current, brand-name, second-hand pieces.
www.styleeyez.com or 403-370-9701

Rewind Consignment Clothing offers clothing that has passed the quality test by having already been worn and washed. Find brand-name, vintage and one-of-a-kind pieces.
www.rewindconsignment.com or 403-263-6669

Trend Clothing Company sells gently used, brand-name clothing, shoes and accessories from their store in stylish Kensington. Voted FFWD *Weekly*'s 2008 Best Consignment/Vintage Clothing Store.
403-283-1167

Vespucci Consignment carries everything from casual wear to high-end designer labels. Its 8,000-square-foot Calgary location makes it the largest ladies' consignment store in the city.
www.vespucci2006.ca or 403-252-9558

3. NATURAL COSMETICS AND PERSONAL CARE PRODUCTS

2bu creates natural, aromatherapy-based skin and personal care products that are free of fillers, synthetic preservatives, parabens and paraffins, propylene glycol and artificial colour and fragrance. They create the products on order to ensure the ingredients are fresh, including therapeutic essential oils. Custom-blending of oils can be done to suit individual skin types and conditions.
www.2bu.ca or 403-478-4178

All Things Jill is a Calgary-based company that has a line of completely natural baby, bath and body products.
www.allthingsjill.ca or 403-982-2852

The Beehive makes its products by hand here in Calgary – soaps, creams, balms, ointments, bath and body care – with natural ingredients like honey and even Bernard Callebaut chocolate!
www.thebeehiveonline.com or 403-270-2622

Eden Essentials at the Calgary Farmers Market carries natural beauty products, lots of which are made locally.
403-271-2078

Krista's Naturals wants their clients to be confident that they are getting only natural, non-toxic skin care, bath and makeup products. All products are handmade in Calgary, and there is nothing you can't pronounce on their ingredients list. Best of all, they smell and feel terrific.
www.kristasnaturals.com or 403-441-6512

Rocky Mountain Soap Company creates handmade soap, skin care, bath and aromatherapy products from natural ingredients such as essential oils, grains and berries.
www.rockymountainsoap.com or 1-877-679-2214

Saje is a line of health, beauty and personal care products that is free of petrochemicals, preservatives, artificial colours, synthetics and animal products.
www.saje.ca or 403-451-7195

> **Green Fact**
>
> In 2003, the European Union banned phthalates in cosmetics sold in Europe, but American and Canadian regulators have not followed suit. Phthalates are a family of chemicals used because they cling to the skin and nails to give products like hairspray, deodorant, nail polish and perfume more staying power. They have been shown to damage the liver, kidneys, lungs and reproductive systems in animal studies.

RESOURCES

1. GUIDES AND RANKINGS

The non-profit Environmental Working Group (EWG) offers free online access to its Skin Deep database, which evaluates more than 47,000 cosmetics and beauty products across 50 definitive toxicity and regulatory databases. How do your cosmetics rank?
www.cosmeticsdatabase.com

Environmental Working Group also has a handy wallet guide that tells you what ingredients to avoid.
www.ewg.org/files/EWG_cosmeticsguide.pdf

Guide to Less Toxic Products is a Canadian source for information regarding the risks and alternatives to common personal care items.
www.lesstoxicguide.ca (choose Personal Care)

Want to know the potential health effects of toxic ingredients in your personal care products? Check out *National Geographic*'s Green Guide Label Decoder.
www.thegreenguide.com/health-safety/dirty-dozen-decoder

ENTERTAINMENT AND RECREATION

GET ON THE GREEN TEAM

We can make choices in our leisure time that are more environmentally responsible. Instead of taking a Sunday drive, take a Sunday stroll. Use a canoe instead of a motorboat to explore local waterways. Ride your bike to a friend's for a Friday movie night instead of driving there.

If you do own motorized recreational vehicles – such as motorboats, motorcycles or quads – keep them maintained much as you would your car. Change the oil and check tire pressure regularly to ensure they are running at peak performance for maximum fuel efficiency.

By supporting Calgary-owned restaurants that are proud to provide food that is local, whole and organic, we support local business people, farmers and producers. This keeps our dollars in the community and keeps the miles our food and drink travel low.

PRODUCTS AND SERVICES IN CALGARY

1. CARDS AND PARTY TIPS

XO Cards has a line of handmade cards made from recycled materials called ECO XO. Also check out their reusable organic cotton gift bags, eco entertaining tips and eco party packs.
www.xocards.ca or 403-265-5455

2. CATERING WITH A CONSCIENCE

Infuse Catering is a full-service catering company that buys at least 70 per cent of its food directly from Albertan farmers who practise sustainable farming methods. Their menu is created around these local foods. Furthermore, they green their operations by composting organics and using environmentally responsible cleaning products.
www.infusecatering.com or 403-269-3902

Red Tree Catering is passionate about creating great food using ingredients from local farms. About 50 per cent of their ingredients are local or organic.
www.redtreecatering.com or 403-242-3246

3. DINING OUT, SUSTAINABLY

A Ladybug Bakery & Café now has two locations, where pastry chefs create decadent treats using organic flour, eggs and milk. The new café location on Aspen Stone Boulevard sw serves soups, sandwiches and panini made from local meats and vegetables. The original location is at the Calgary Farmers Market.
403-287-1137

Broken City supports the local economy by allowing people to pay 50 per cent of their tab with Calgary Dollars – Calgary's local currency – before 9 p.m. Their environmental commitment includes recycling bottles and cardboard, composting food waste and sourcing local food and drink. For example, their draft beers are Calgary-brewed Big Rock and Wild Rose.
www.brokencity.ca or 403-262-9976

The Coup and its sister café/lounge Meet provide vibrant atmosphere and creative vegetarian fare while promoting local whole and organic foods. They are committed to environmental sustainability, composting organic material and using leftover cooking oil to fuel a vehicle.
www.thecoup.ca or 403-541-1041

Convergence Café created their menu around local products that would be available year round. Things you won't find here are big brand-name sodas, deep fryers or microwave ovens. But you will find organic, locally roasted coffee, natural cleaning products made in-house and a recycling system.
www.convergencecafe.com or 403-206-1564

Cuisine Concepts has several venues under its umbrella: Diner Deluxe, Big Fish, Open Range, Urban Baker and Vue Café. A third party picks up their compost, milk containers, plastic, cardboard and paper. Diverting compost from their waste stream has made such a big difference that they have been able to downsize their garbage container and frequency of garbage pick-up at the Open Range and Big Fish locations. They also use local food first. Diner Deluxe has its own farmer, Farmer Cliff, and they buy all his potatoes, eggs, chickens and vegetables.
www.cuisineconcepts.ca

FARM uses products from local growers and producers, which allows them to know the farmers and have confidence in the products they serve. They also have a composting and recycling program to reduce the waste produced by the restaurant.
www.farm-restaurant.com or 403-245-2276

Good Earth Café, started in Calgary in 1991, now has 13 Calgary locations. They are committed to making environmentally and socially responsible decisions about the coffee they serve: 100 per cent shade grown and organic, roasted by a family-run company in Calgary. As of April 2008, all their cafés are wind powered, offsetting more than 500 tonnes of carbon dioxide emissions annually.
www.goodearthcafes.com

Ground Effect Café makes almost everything they serve entirely from scratch with locally grown products. They employ vermicomposting and traditional composting systems, which means they are nearly waste-free: they don't even have garbage pick-up.
6622 - 20A Street SE or 403-236-9003

Higher Ground describes itself as an organic coffee house. All of their coffee is organic, shade grown and fairly traded. The loose teas they serve are also organic.
www.highergroundcafe.ca or 403-270-3780

Jugo Juice has been reducing the waste from their disposable cups. After experimenting with corn-based compostable cups, they opted for #1 PET plastic cups made of 20 per cent post-consumer recycled plastic. These cups are also recyclable, unlike Styrofoam.
www.jugojuice.com

River Café changes their menu monthly to take advantage of food that is in season. They use organic and sustainably harvested food from local producers and their own organic garden. They even preserve their own produce by canning and freezing it for use over the winter.
www.river-cafe.com or 403-261-7670

Rouge has their own on-site vegetable and herb gardens so they can serve meals using the freshest local ingredients.
www.rougecalgary.com or 403-531-2767

Wild Rose Brewery not only makes local beer, but they also donate the leftover grain from the brewing process to a local farmer who feeds it to his cattle. Also, their taproom is furnished with repurposed items.
www.wildrosebrewery.com or 403-720-2733

4. EVENT PLANNING

Clean Calgary Association has a *Green Events Guide* to help you minimize the waste generated during your event.
www.cleancalgary.org (under Guides choose *Green Events Guide*)

Mingle Event Management creates events with environmental responsibility in mind. They establish environmental goals and priorities right from the start to deliver events that are good for business and the environment.
www.minglemyevent.com or 403-454-3832

5. ENJOYING THE OUTDOORS

Mountain Equipment Co-op encourages people to recycle outdoor gear with their free Online Gear Swap.
www.mec.ca

RESOURCES

1. FILMED ENTERTAINMENT, WITH A GREEN PURPOSE

The Great Warming, narrated by Alanis Morissette and Keanu Reeves, reveals how climate change is affecting the lives of people the world over. Produced by a Canadian company, the documentary is based on the book *Storm Warning: Gambling with the Climate of Our Planet*, by Canadian science writer Lydia Dotto.
www.thegreatwarming.com

Radiant City, by Calgary native Gary Burns, is a film about suburban sprawl. "Sprawl is eating the planet. Across the continent the landscape is being levelled – blasted clean of distinctive features and overlaid with zombie monoculture. Politicians call it growth. Developers call it business. The Moss family call it home."
www.radiantcitymovie.com

2. GREEN LIFESTYLE GUIDES

Dine Alberta promotes quality, locally grown food in union with the culture, geography and history of Alberta. On their website, you can find restaurants that feature Alberta regional cuisine and samples of their delectable regional menus.
www.dinealberta.ca

Green Living is a Canadian magazine whose goal is to show readers how to live green with their home, beauty and lifestyle choices. See their online Green Living Guide for Calgary that highlights local businesses that offer the best green alternatives.
www.greenlivingonline.com

Happy Cow and VegDining.com have online listings of vegetarian restaurants in Calgary and around the world.
www.happycow.net
www.vegdining.com

GET GREENER IN THE WORKPLACE

GET ON THE GREEN TEAM

People who make greener choices at home also want to make responsible choices at work. Furthermore, businesses are realizing that their green reputation counts. Here are some ideas on how to keep it green at the office:

- Recycled paper for your printer and copier is now widely available at a reasonable price.
- Every sheet of paper has two sides; use them both! But try to keep the printing to a minimum by keeping digital files instead of paper.
- Consider second-hand office furniture.
- Turn computers off when not in use. Using a powerbar with an on/off switch allows you to shut off all equipment, including printers and scanners, and eliminates phantom loads.
- Ensure power management options are enabled on your computer, which can result in energy savings of 60 to 80 per cent.
- Get rid of paper and Styrofoam cups... completely! Use reusable mugs, glasses and dishes.
- Remember to e-cycle your computers and other electronics when you replace them.
- Bring litterless lunches to work.

PRODUCTS AND SERVICES IN CALGARY

1. BANKING

Citizens Bank has an ethical policy that helps them decide which enterprises they want to do business with. They offer environmentally sensitive banking choices, such as a VISA card that donates to Amnesty International or Oxfam every time you use it. They are 100 per cent Canadian owned and have an ethical policy stating that they seek to do business with organizations that demonstrate a commitment to environmental leadership and ethical business practices.
www.citizensbank.ca or 1-888-708-7800

Climate Friendly Banking calculates the carbon footprint of your account with Canada's big banks, and suggests alternatives.
www.climatefriendlybanking.org

2. CARBON OFFSETTING

Does your business involve a lot of air travel? Do you feel that your office's carbon footprint, even after intense scrutiny and cutting back, is still too big? Carbon Offset Solutions has information on all things related to emissions trading and climate change policy in Canada. Find out how to measure greenhouse gas offsets, a catalogue of emissions trading resources, an offset registry and more.
www.carbonoffsetsolutions.ca or 403-517-2700

3. CARPOOLING AND COURIER SERVICES

Earth's Courier runs its company vehicles using bio-diesel made from recycled restaurant grease, reducing their greenhouse gas emissions by 75 per cent.
www.earthscourier.com or 403-276-4646

Namaste Messenger is an employee-owned, Calgary-based shipping company dedicated to sustainable business practices. Their goal is

to cover 80 per cent of the city by bicycle. The motto: live for today while building for tomorrow.
www.namastemessenger.ca or 403-477-3737

Sustainable Alberta Association provides information, technical assistance and incentives for employers and employees to reduce the number of vehicles on the road.
www.calgarycommute.ca or 403-294-0904

4. CONSULTING SERVICES, GUIDES AND WORKSHOPS

The Canada Green Building Council Alberta Chapter seeks to transform the built environment so it is environmentally responsible, profitable and healthy. Their mandate includes homes, commercial and industrial buildings, and entire neighbourhoods. They do this through programs like the LEED Canada Initiative and the Green Building Performance Initiative, through which builders can certify their buildings.
www.cagbc.org/chapters/alberta or 780-669-3665

Clean Calgary Association delivers a 50-minute presentation called "Waste at Work" to help employees reduce, reuse and recycle materials at work.
www.cleancalgary.org (under Education click on Businesses and Organizations) or 403-230-1443

Commercial Environmental Services empowers companies to reduce, reduce and recycle environmental waste. They provide waste consulting services, an online recycling directory and the online Calgary Materials Exchange, where companies can list materials that are available or wanted for exchange.
www.cleancalgary.org (under Programs choose Commercial Environmental Services) or 403-230-1443

Green Fuse Communications specializes in writing, event management and workshops. They focus on projects that reflect their passion for environmentally sound and sustainable lifestyles.
www.greenfuse.ca or 403-244-2211

REAP (Respect for the Earth and All People) is a not-for-profit association of Calgary businesses that are concerned with issues of sustainability. Through their free online magazine and events they provide information and inspiration to help Calgarians make choices that have positive consequences for the Earth and all people.
www.reapcalgary.com or 403-862-2874

5. ENVIRONMENTALLY RESPONSIBLE CLEANING SERVICES

Skylark Cleaning Services provides commercial and residential cleaning services that use only green cleaning products in order to protect the environment and people's health.
403-978-1611

6. GREEN CAREERS

Environmental Careers Organization (ECO) Canada helps individuals build meaningful environmental career and provides employers with resources to find and keep the best environmental practitioners.
www.eco.ca or 403-233-0748

Young Environmental Professionals promotes opportunities for emerging environmental and sustainability professionals.
www.yepcanada.ca/calgary

7. GREEN COMPUTER SERVICES

Tech House Computer Services provides on-site computer services to business and residential users in an environmentally responsible manner. They are aware that the small choices they make can create big change. That's why they donate used equipment to the Calgary Drop In Centre, use 100 per cent recycled paper and are members of REAP (see above).
www.techhouse.ca or 403-616-2519

8. LIGHTS ON!

Calgary Lighting Products provides commercial and industrial lighting solutions, and they have their own onsite recycling machine for fluorescent lighting tubes to provide safe and responsible disposal for their clients.
www.calgarylightingproducts.com or 403-258-2988

9. SUSTAINABLE OFFICE INTERIORS

ELEMENT Integrated Workplace Solutions Inc. specializes in sustainable modular office interiors that can be reused and reconfigured in a variety of formats. They carry products that are GREENGUARD and Forest Stewardship Council certified, and they can help companies achieve LEED certification. They are the southern Alberta distributor for DIRRT, which manufactures flexible and sustainable workspaces right here in Calgary: movable walls and access floors that virtually eliminate construction waste and any future renovation waste, and provide a productive, long-term, flexible asset.
www.elementiws.com or 403-444-7390
www.dirtt.net or 403-723-5040

10. GREEN OFFICE SUPPLIES

Clean Calgary Association's The EcoStore has the best selection of environmentally friendly office supplies in town. Find eco-friendly versions of the staples: paper, envelopes and file folders. But also find biodegradable pens, recycled paperclips, recycled plastic clipboards and much more.
www.cleancalgary.org or 403-230-1443

11. RECYCLING SERVICES

Buy remanufactured toner cartridges and refill ink cartridges for your printers and fax machines. Find local dealers in The City of Calgary's *Residential Recycling Opportunities* guide.

www.calgary.ca (search "recycling opportunities," choose Recycling Information, and under downloads, click Recycling Opportunities List; look at the section Reusable Printer and Toner Cartridges/Ribbons)

GEEP Ecosys Inc. is different from other recyclers because it starts by *reusing* eligible equipment. They remarket quality computer equipment that has come from commercial clients – with permission and after wiping all data – at their store in the northeast. Anything that can't be reused is disassembled and recycled.
www.geepecosys.com or 403-219-3137

Plastics recycling for commercial and industrial services can be found at:

- Merlin Plastics Alberta Inc.
  www.merlinplastics.com or 403-259-6637
- Plastic Resources International
  403- 236-4485
- PEL Recycling
  www.calgaryrecycles.com or 403-369-3777

WhyWasteIt.ca is a waste diversion and recycling directory for Alberta's construction and demolition industries. It is sponsored by EnerVision, whose mission is to help the home building industry get greener.
www.whywasteit.ca or 403-705-9007

12. WORKPLACE EVENTS

Clean Calgary's *Green Events Guide* can help you reduce waste at your next event.
www.cleancalgary.org (under Guides)

Mingle Event Management creates sustainable events by identifying green goals at the beginning of the planning process. In the end, clients receive a sustainability audit to highlight exactly how much energy and resources they've saved by choosing Mingle Event Management to plan their events.
www.minglemyevent.com or 403-454-3832

RESOURCES

1. GUIDES, TIPS AND LISTS

Canadian Centre for Pollution Prevention has tools to help small businesses improve their environmental performance.
www.c2p2online.com (under Affiliated Websites choose P2 [Pollution Prevention] for Small Business)

Corporate Knights is Canada's magazine for responsible business. Its content focuses on how companies impact people and the planet, and each June issue includes the annual list of Canada's Best 50 Corporate Citizens.
www.corporateknights.ca

The Government of Canada's *Green Meeting Guide* has lots of suggestions about how to green your meeting, from transportation and accommodation to office procedures and selecting a sponsor.
www.greeninggovernment.gc.ca (under In Focus)

Guide to Greener Electronics by Greenpeace rates the top manufacturers of electronics such as personal computers, mobile phones, TVs and games consoles according to those manufacturers' publicly announced policies on toxic chemicals, recycling and climate change.
www.greenpeace.org (search "guide to greener electronics")

Review the cleaning products used in your workplace using the Cleaners and Toxins Guide, a free pdf download from Toxic Free Canada.
www.toxicfreecanada.ca

2. TRAINING AND CERTIFICATION

BOMA (Building Owners & Managers Association) Best is a national environmental certification for commercial buildings. It assesses how well buildings are performing environmentally and provides tools for owners to reduce energy consumption and improve waste management.
www.bomabest.com or 403-237-0559

Consider obtaining LEED™ (Leadership in Energy & Environmental Design) accreditation on your next commercial building or renovation. Source LEED Accredited Professionals for your design and construction teams. They understand green building practices and know how to get your project LEED certified.
www.cagbc.org or 1-866-941-1184

The Natural Step Canada is a non-profit organization that provides training and advice to businesses on how to integrate sustainability principles into their operations. They use a specific, science-based definition of sustainability and focus on practicality and results.
www.naturalstep.ca

REBATES AND GOVERNMENT PROGRAMS

1. THE CITY OF CALGARY WATER SERVICES

The City of Calgary Water Services department offers programs to help industrial, commercial and institutional customers reduce water consumption.
www.calgary.ca (in the A-Z Directory choose W, then Water & Wastewater, then Commercial Water Efficiency)

2. NATURAL RESOURCES CANADA

Natural Resources Canada provides information, tools and incentives for businesses that are retrofitting commercial buildings, improving fuel management of fleet vehicles and reducing energy used by commercial and industrial equipment. You will also find information on financial assistance for these ventures.
www.oee.nrcan.gc.ca (under For Business Use)

3. OFFICE OF ENERGY EFFICIENCY

The Office of Energy Efficiency's *Energy Savings Toolbox* provides a step-by-step methodology to help industry identify and capitalize on energy savings. It will help you evaluate and manage energy use in every phase of an operation, from designing and conducting an energy audit to carrying out cost-benefit analyses.

www.oee.nrcan.gc.ca (under For Business Use, choose Industrial: Facilities and Equipment, then Technical Information, then Newsletters, Case Studies and Technical Guides. Scroll down to Technical Guides.)

> **Green Fact**
> Laptops use less energy than desktops, and flat-screen LCD monitors use 70 per cent less energy than standard monitors. They also contain 95 per cent less lead, which is a toxic heavy metal.

15
GENERAL RESOURCES FOR LIVING GREEN

GET ON THE GREEN TEAM

There are many organizations that are willing and able to help you green your life.

PRODUCTS AND SERVICES IN CALGARY

1. CLEAN CALGARY ASSOCIATION

Clean Calgary Association's mission is to empower Calgarians to create healthy homes and communities by providing environmental education, products and services. The EcoStore carries a wide variety of environmentally friendly products; they have many guides that answer common questions; and they hold classes to help you make more environmentally responsible choices. Clean Calgary is a key resource for all Calgarians interested in living a more environmentally friendly lifestyle.
www.cleancalgary.org or 403-230-1443

2. GUIDES, WORKSHOPS AND PLANS

Conscious Home provides green lifestyle coaching for forward-thinking people who want to create and sustain a more positive lifestyle. Services range from one-on-one in-home consultations to corporate presentations, home parties and workshops.
www.conscioushome.ca or 403-475-7688

EarthWise Solutions creates Green Plans to help homeowners and condominium boards consume less energy, produce less

waste and save money. These customized plans save hours of time and make your space and the planet healthier. For corporate clients they create and deliver customized presentations on green living issues.
www.earthwisesolutions.ca or 403-896-5096

The Green Guide appears every Friday in the *Calgary Herald*'s "Real Life" section and is the place to go to keep your thumb on the pulse of what is green in Calgary.
www.calgaryherald.com/life/green-guide/index.html

Green Living is a Canadian magazine with the goal of showing readers how to live green with their home, beauty and lifestyle choices. See their online *Green Living Guide for Calgary* that highlights local businesses that offer the best green alternatives.
www.greenlivingonline.com

Simply Going Green in Three Years or Less by local author Kaayla Canfield helps you create a plan to implement green projects into your home and life, as well as accomplish your green goals.
www.simplygoinggreen.ca

Sustainable Calgary has created a *Green Map for Calgary*. It highlights environmentally important resources in your neighbourhood: natural areas, landfills, recycling depots and much more. They also provide support for community action toward a sustainable future in Calgary through programs like The Citizens' Agenda.
www.sustainablecalgary.ca (click the Green Map icon)

3. TAKING ACTION

Green Drinks is a monthly gathering of people who are concerned about the environment in Calgary. The group brings together like-minded people to socialize and network.
www.greendrinkscalgary.ca

The Sierra Club of Canada's mission is to develop a diverse, well-trained grassroots network working to protect the integrity of our local and global ecosystems. The Chinook chapter is specifically committed to protecting and improving the environment and ecosystems of Calgary and Southern Alberta.
www.sierraclubchinook.org or 403-233-7332

Step It Up Alberta seeks action against climate change from governments, corporations and institutions. They are not associated with any political party and welcome new members to help them raise public awareness about environmental issues and ask the government to invest in alternative energy technologies.
www.stepitupalberta.ca

RESOURCES

1. BOOKS

Ecoholic, by Adria Vasil, is a Canadian book of practical tips and products for living a little greener.
www.ecoholic.ca

The Geography of Hope, by Calgary author Chris Turner, examines solutions for a sustainable future that are already at work around the world.
www.thegeographyofhope.com

2. CALGARY DOLLARS

The Arusha Centre's Calgary Dollars is a grassroots local currency. It encourages the consumption of local products and services to strengthen our local economy and build community.
www.calgarydollars.ca

3. CERTIFICATION BODIES AND THEIR LOGOS

The Canada Organic logo identifies agricultural products containing at least 95 per cent organic ingredients. This federal certification pro-

gram is administered by the Canadian Food Inspection Agency.
www4.agr.gc.ca (search "Biologique Canada Organic")

EcoLogo^M is owned by Environment Canada. Its purpose is to help consumers and industry make more environmentally conscious decisions. Many products are EcoLogo-certified, including cleaners, detergents, paper products, fuels, building supplies and more.
www.ecologo.org

EnerGuide and ENERGY STAR are initiatives of Natural Resources Canada that help consumers purchase the most energy-efficient electronics and appliances. The EnerGuide label gives an estimate of the annual energy consumption of an appliance and ranks it against other appliances of the same size and type. The ENERGY STAR symbol identifies the most energy-efficient products on the market.
www.oee.nrcan.gc.ca/energuide
www.oee.nrcan.gc.ca/energystar

What are the meaningful eco-labels in Canada? Find out from Environment Canada, and even download a handy wallet guide that you can refer to when shopping.
www.ec.gc.ca/education (in the left-hand sidebar, click What You Can Do, then, under At home, choose Buying green products, then Green buying: Guide to ecolabels)

Forest Certification recognizing sustainable forest management in Canada is recognized by three systems:

- Canadian Standards Association (CSA) (PEFC Canada)
  www.csa-international.org (choose Product Areas, then Sustainable Forest Management Program)
- Forest Stewardship Council (FSC)
  www.fsccanada.org
- Sustainable Forestry Initiative® (SFI)
  www.sfiprogram.org

TransFair is Canada's certifying body for fairly trade goods. In Canada, fairly traded bananas, cocoa products, coffee, cotton, flowers, rice,

sugar, tea and more is marked with the TransFair logo. Their website tells you more about certified products and where to find them.
www.transfair.ca

4. CONSULTING SERVICES

Ravis Sustainable Consulting offers courses and consulting services to help people navigate the complex arena of energy, conservation, green technology and sustainability. They are also experts in permaculture.
www.ravissustainable.com or 403-770-9789

5. EVENTS

EcoLiving™ Events Ltd. facilitates information exchange and partnerships among communities, consumers and businesses about environmentally sustainable options. They host the EcoLiving Fair annually to highlight green products and resources, and the EcoLiving Tour, which showcases Calgary area buildings that have incorporated practical solutions to reduce energy costs.
www.ecolivingfair.ca

6. HANDY WEBSITES

Climate Change Central is a public/private partnership promoting innovative responses to global climate change and its impacts in Alberta. It also administers provincial rebate programs, such as those for adding insulation, purchasing an energy-efficient washing machine or upgrading your furnace.
www.climatechangecentral.com

The Eco-Footprint Exchange from The City of Calgary is a database of information, programs and services in Calgary that you can use to reduce your ecological footprint.
www.calgary.ca (search "eco footprint exchange")

> **Eco-words**
> *Ecological footprint:* the amount of land and water a group of people uses to produce the resources it consumes and to absorb its waste.

> **Green Fact**
> The average Calgarian's ecological footprint is 9.86 hectares. The Canadian average is 7.25 hectares. Canadians have the fourth largest global ecological footprint in the world; the global average is 2.7 hectares. That means Calgarians live some of the least sustainable lifestyles on the planet. See find out more about Calgary's footprint.

Environment Canada has resources to help you go green on the road, at home, at work, at school and in the community.
www.ec.gc.ca/education (under Take Action for the Environment choose What You Can Do)

The Office of Energy Efficiency's website is the hub for the Government of Canada;s energy conservation programs, including the ecoENERGY Retrofit grants for homeowners and businesses. This website is filled with information aimed at helping Canadians save money while contributing to a healthier environment.
www.oee.nrcan.gc.ca

The Pembina Institute has created One Less Tonne, an online tool to help Canadians choose actions for cutting their greenhouse gas emissions and see the money they can save in energy costs.
www.onelesstonne.ca

David Suzuki's website is full of resources to help you conserve nature and improve your quality of life.
www.davidsuzuki.org

REBATES AND GOVERNMENT PROGRAMS

1. ONE-STOP MONEY-SAVING DATABASE

Environment Canada has a one-stop shopping database to search for grants, rebates, discounts and other incentives to help you live a more environmentally friendly lifestyle.
www.incentivesandrebates.ca

APPENDIX

COMMUNITY RECYCLING DEPOT LOCATIONS 2009

NORTHWEST NEIGHBOURHOOD DEPOTS
Beacon Hill 11320 Sarcee Tr. N.W. (West side of Home Depot parking lot)
Beddington 8120 Beddington Blvd. N.W. (Northeast corner of Safeway parking lot)
Bowness 7937 43 Ave. N.W. (North side of old Safeway parking lot)
Confederation Park 2807 10 St. N.W. (North side of Rosemont Community Assn. parking lot)
Crowfoot 90 Crowfoot Way N.W. (South side of Rona parking lot)
Dalhousie Station 5005 Dalhousie Dr. N.W. (Southeast side of parking lot)
Hamptons 1000 Hamptons Dr. N.W. (North side of Co-op parking lot)
Hillhurst/Sunnyside 1320 5 Ave. N.W. (East side of community assn. parking lot)
Huntington Hills 7040 4 St. N.W. (Superstore parking lot)
Market Mall 3625 Shaganappi Tr. N.W. (Northeast corner of parking lot)
North Hill 1901 16 Ave. N.W. (East side of mall parking lot)
Silver Springs 5720 Silver Springs Blvd. N.W. Tuscany 5019 Nose Hill Dr. N.W. (Northeast side of Home Depot parking lot)
Varsity Varley Dr. N.W. (Southeast of 32 Ave. & 39 St. N.W.)

NORTHEAST NEIGHBOURHOOD DEPOTS
Bridgeland Rehabilitation Society 7 11 St. N.E. (East side of the Rehabilitation Society parking lot)
Castleridge 55 Castleridge Blvd. N.E. (North side of Safeway parking lot)
Country Hills 11622 Harvest Hills Blvd. N.E. (North side of North Pointe Park 'n' Ride)
Horizon 3550 32 Ave. N.E. (Southeast corner of Safeway parking lot)
Marlborough 500 Block of 36 St. N.E. (West side of mall parking lot, along 36 St. N.E.)
Monterey 2220 68 St. N.E. (Southwest corner of Co-op parking lot)
Renfrew 16 Ave. & Russet Rd. N.E. (Renfrew Athletic Park)
Sunridge 2525 36 St. N.E. (Southeast corner of mall parking lot)
Trans Canada East 300, 1440 52 St. N.E. (South side of Safeway)
Village Square 2520 52 St. N.E. (South side of Co-op parking lot)

SOUTHWEST NEIGHBOURHOOD DEPOTS
Bridlewood 2335 162 Ave. S.W. (Behind medical clinic)
Chinook 6455 Macleod Tr. S.W. (Southeast corner of Chinook Shopping Centre parking lot)
Connaught 813 11 Ave. S.W. (East side of Safeway)
Glenmore Landing 1600 90 Ave. S.W. (Northeast corner of Safeway parking lot)
Lakeview 6449 Crowchild Tr. S.W. (East side of Sobeys parking lot)
Millrise 150 Millrise Blvd. S.W. (East side of Sobeys parking lot)
Oakridge 2580 Southland Dr. S.W. (South side of Co-op parking lot)
Richmond Square 3915 51 St. S.W. (North side of parking lot)
Scarboro 15 Ave. & 16 St. S.W. (Southwest side of Tennis Club)
South Calgary 3130 16 St. S.W. (community assn. parking lot)
Spruce Cliff 3400 block of Spruce Dr. S.W. Strathcona 555 Strathcona Blvd. S.W. (Near Sobeys)
Trans Canada West Southwest of intersection of Stoney Tr. North and Trans Canada Hwy. West
West Hills 200 Stewart Grn. S.W. (Northwest corner of parking lot behind GAP)

SOUTHEAST NEIGHBOURHOOD DEPOTS
Acadia 383 Heritage Dr. S.E. (Southeast corner of Acadia Shopping Centre)
Deerfoot Meadows 100, 20 Heritage Meadows Way S.E. (Northwest corner of Superstore parking lot)
Deer Valley 1221 Canyon Meadows Dr. S.E.(North side of mall parking lot)
Douglasdale 11520 24 St. S.E. (Behind Sobeys)
Forest Lawn 3330 17 Ave. S.E. (Northeast corner of Co-op parking lot)
Macleod 9630 Macleod Tr. S.E. (North side of Rona parking lot)

Manchester Dartmouth Rd. (between 30 Ave and 34 Ave.)
McKenzie Towne 20 McKenzie Towne Ave. S.E. (Southeast side of Sobeys parking lot)
Midnapore Sundance Bannister Rd. & 153 Ave. S.E. (Beside Centrex gas station/carwash)
Ogden 7740 18 St. S.E. (West side of Safeway parking lot)
Shawnessy 390 Shawville Blvd. S.E. (West side of Home Depot parking lot)
Southcentre 11011 Bonaventure Dr. S.E. (Northeast side of Safeway parking lot)
South Trail Depot 5125 126 Ave. S.E. (Southwest corner of Home Depot parking lot)
Village Square 2520 52 St. N.E. (South side of Co-op parking lot)
West Hills 200 Stewart Grn. S.W. (Northwest corner of parking lot behind GAP)

ELECTRONICS RECYCLING DEPOT LOCATIONS 2009

NORTHWEST
Market Mall Staples – eCycle Solutions 3625 Shaganappi Tr. N.W.
Best Buy Northland – eCycle Solutions 5111 Northland Dr. N.W.
Future Shop Northland – eCycle Solutions 420, 5111 Northland Dr. N.W.
Future Shop Beacon Hill – eCycle Solutions 11810 Sarcee Tr. N.W.

NORTHEAST
TopFlight 3121 16 St. N.E.
Northgate Staples – eCycle Solutions 121, 565 36 St. N.E.
32nd Avenue N.E. Staples – eCycle Solutions 3030 32 Ave. N.E.
Coventry Hills Staples- eCycle Solutions 130 Country Village Rd. N.E.
GEEP Ecosys Inc. 950 64 Ave. N.E
Best Buy Sunridge – eCycle Solutions 500, 3221 Sunridge Way N.E.
Future Shop Sunridge – eCycle Solutions 3319 26th Ave. N.E.
Future Shop Coventry Hills – eCycle Solutions 331 130 Country Village Rd. N.E.

SOUTHWEST
Recycle-Logic Inc. 4324 Quesnay Wood Dr. S.W.
Signal Hills Staples - eCycle Solutions 5662 Signal Hill Centre Dr. S.W.
Chinook Staples - eCycle Solutions 321 61 Ave. S.W.
City Centre Staples - eCycle Solutions 1215 9th Ave. S.W.

SOUTHEAST
Southtrail Crossing Staples - eCycle Solutions 4307 130 Ave. S.E.
Ecco Waste Landfill Systems - eCycle Solutions 9908 24 St. S.E.
BFI Landfill - eCycle Solutions 201 194 Ave. S.E.
Foothills Staples - eCycle Solutions 100, 3619 61 Ave. S.E.
Shawnessy Staples – eCycle Solutions 140, 350R Shawville Blvd. S.E.
Best Buy Deerfoot Meadows – eCycle Solutions 300, 8180 11 St. N.E.
Future Shop Deerfoot Meadows – eCycle Solutions 1180, 33 Heritage Meadows Way S.E.
Future Shop Shawnessy – eCycle Solutions 110, 350R Shawville Blvd. S.E.

HOUSEHOLD CHEMICAL DROP-OFF LOCATIONS 2009

#4 Fire Station 1991 18 Ave. N.E.
#17 Fire Station 3740 32 Ave. N.W.
#20 Fire Station 2800 Peacekeepers Way S.W.
#24 Fire Station 2607 106 Ave. S.W.
#26 Fire Station 450 Midpark Way S.E.
East Calgary Landfill 68 St. & 17 Ave. S.E.
Shepard Landfill 52 St. & 114 Ave. S.E.
Spyhill Landfill 69 St. & 112 Ave. N.W.